XVI International Symposium on Electron Molecule Collisions and Swarms 2009

Journal of Physics: Conference Series Volume 204

Toronto, Ontario, Canada
29 July - 1 August 2009

ISBN: 978-1-61738-273-4

Printed from e-media with permission by:

Curran Associates, Inc.
57 Morehouse Lane
Red Hook, NY 12571

Some format issues inherent in the e-media version may also appear in this print version.

Printed by Curran Associates, Inc. (2010)

For permission requests, please contact the Institute of Physics
at the address below.

Institute of Physics
Dirac House, Temple Back
Bristol BS1 6BE UK

Phone: 44 1 17 929 7481
Fax: 44 1 17 920 0979

techtracking@iop.org

Additional copies of this publication are available from:

Curran Associates, Inc.
57 Morehouse Lane
Red Hook, NY 12571 USA
Phone: 845-758-0400
Fax: 845-758-2634
Email: curran@proceedings.com
Web: www.proceedings.com

TABLE OF CONTENTS

PAPERS

XVI International Symposium on Electron Molecule Collisions and Swarms IOP Publishing
Journal of Physics: Conference Series **204** (2010) 012001 doi:10.1088/1742-6596/204/1/012001

Electron-driven excitation and dissociation of molecules

S. Chourou[1], Å. Larson[2] and A. E. Orel[1]

[1]Department of Applied Science, University of California, Davis, USA
[2] Dept. of Physics, Stockholm University, AlbaNova University Center S-106 91 Stockholm, Sweden

E-mail: aeorel@ucdavis.edu

Abstract. In the collision of electrons with molecules and molecular ions, excitation and dissociation are dominated by resonant processes, where the electron becomes temporarily trapped, changing the forces felt by the nuclei. In this paper, we will outline our method for treating these collision processes, where one or more resonant states exist. We separate the problem into two steps. First we carry out *ab initio* electron scattering calculations at fixed internuclear geometries to determine the resonant energy surfaces and the corresponding surface of autoionization widths, using the Complex Kohn variational method. These resonance positions and widths are then used as input to a dynamics study to determine the cross section and product distributions for the dissociation or excitation process. We will present results on a number of systems, including HCCH, HCN/HNC and HCCCN as examples of dissociative attachment, and H_2O^+ for dissociative recombination.

1. Introduction

Electron collisions with molecules and molecular ions that lead to dissociation play a key role in a number of environments, since they produce the radicals and molecular fragments that initiate and drive the relevant chemistries. Examples of such systems can be found in biology as well as in chemistry and physics; they range from the technologically important plasmas used in plasma enhanced chemical vapor deposition [1], to planetary atmospheres and interstellar clouds [2]. Due to the difference between the mass of the electron and the nuclei, it is unlikely that there is a significant cross section for vibrational energy transfer or dissociation in electron-molecule or electron-molecular ion collisions. Nuclei move slowly compared to electrons, so the time scale for nuclear motion is very large compared to the time of an electron collision. However, in resonant systems where the electron temporarily attaches to the molecule, the forces felt between its atoms are changed for a period of time comparable to a vibrational period. This can lead to resonant vibrational excitation (VE), resonant dissociative excitation (DE) and dissociative attachment (DA), for neutral targets, or dissociative recombination (DR) in the case of ions. These processes can be studied by considering the capture of the electron into a resonant dissociative state, and then describing the dynamics of the molecule moving on the excited state (resonant) potential energy surface. The dynamics must be able to describe autoionization since the molecule can re-emit the electron. First we carry out *ab initio* electron scattering calculations at fixed internuclear geometries to determine the resonant energy surfaces and the corresponding surface of autoionization widths, using the Complex Kohn variational method [3].

These resonance positions and widths are then used as input to a dynamics study to determine the cross-section and product distributions for the dissociation or excitation process. In the following section we will briefly describe the underlying theory used in these calculations. We will then illustrate the method with a series of examples.

2. Theory

Theoretical treatments of DR, DA, VE and DE are generally based on rigorous resonance scattering theory [4], formulated within the Born-Oppenheimer approximation. The principal result of the formal theory is the so-called nuclear wave equation that governs the nuclear dynamics on the resonance state(s). This is an inhomogeneous wave equation which, due to the presence of a nuclear Green's function that operates on the nuclear coordinates, involves an effective Hamiltonian that is complex, non-local and energy dependent. In many resonant dissociative recombination/attachment problems, the heavy-particle dynamics can be treated within the local complex potential or "boomerang" approach[5]. In this approximation, the nuclear wave equation at total energy E is

$$(E - K_R - V_{res})\xi_\nu = \phi_\nu \tag{1}$$

where K_R is the nuclear kinetic energy operator, V_{res} is the (complex) resonant state potential,

$$V_{res}(\vec{R}) = V_0(\vec{R}) - i\Gamma(\vec{R})/2 \tag{2}$$

and ξ_ν is the nuclear wave function associated with the electronic resonance state. (Note, in this discussion we have used atomic units ($\hbar = m = 1$). The position and width of the resonance that form the resonant potential are E_{res} and Γ respectively. The driving term for the nuclear wave equation, or "entry amplitude", ϕ_ν is defined as

$$\phi_\nu(\vec{R})) = \left(\frac{\Gamma(\vec{R})}{2\pi}\right)^{1/2} \chi_\nu(\vec{R}), \tag{3}$$

where χ_ν is the initial vibrational wave function of the target. The local complex potential model is expressed in Eqs. (1-3) in its original or "boomerang" form, and provides the wave function ξ_ν from which the cross sections can be calculated. The inhomogeneous differential equation given by Eq. (1), can be recast into this time-dependent form, as first shown by McCurdy and Turner [6]. We then proceed by the direct integration of the time-dependent Schrödinger equation. The time propagation must be carried out long enough for the wave packet to reach the asymptotic region of the potential where the autoionization loss has gone to zero. In the examples that follow, the dynamics calculations were carried out using the MultiConfiguration Time-Dependent Hartree (MCTDH) method [7] and the package of codes developed in Heidelberg by the group of Meyer. We have found this method to be extremely efficient in studying these processes.

The cross-sections resulting from the dynamics calculations are only as accurate as the resonance parameters used. We used the complex Kohn variational method [3] to calculate these parameters. The electronic resonance parameters, $V_0(\vec{R})$ and $\Gamma(\vec{R})$ are determined by analyzing the T-matrix from state-of-the-art electron-molecule or electron-molecular ion scattering calculations. For single, isolated resonances, the eigenphase sum is fit to a Breit-Wigner form.

3. Examples: Dissociative Attachment

3.1. HCCH

Recent experiments on DA of acetylene show a strong peak in the 3 eV energy range, with products $HCC^- + H$ [8]. The electron is believed to attach to an antibonding π^* orbital, but

Journal of Physics: Conference Series **204** (2010) 012001 doi:10.1088/1742-6596/204/1/012001

dissociation occurs not on the C-C bond but on the C-H bond . In addition, the fragments have A′ symmetry. These observations are similar to those seen in DA of formic acid, a system we have previously studied [9]. We carried out *ab initio* calculations for elastic electron scattering from acetylene using the complex Kohn variational method at the static exchange level to obtain the resonance surface and autoionization width before the anion crosses the neutral. After it has crossed additional quantum chemistry calculations, at a consistent level, *ie* SCF, were used to map out the rest of the surface. These calculations were done keeping one C-H distance fixed, reducing the problem to three (Jacobi) dimensions, the C-CH distance (r), the distance from the other hydrogen atom to the center of mass (R) and the angle in-between them (θ). Note that since only total DA cross sections, not final state distributions were desired, the form of the anion surface away from the crossing is not relevant. We then carried out MCTDH calculations [7] to track the dissociation dynamics and obtain the dissociative attachment cross sections [10]. These results are shown in Figure 1. The apparent excellent agreement in the case of HCCH at 0K is fortuitous since the experiment was done at room temperature.

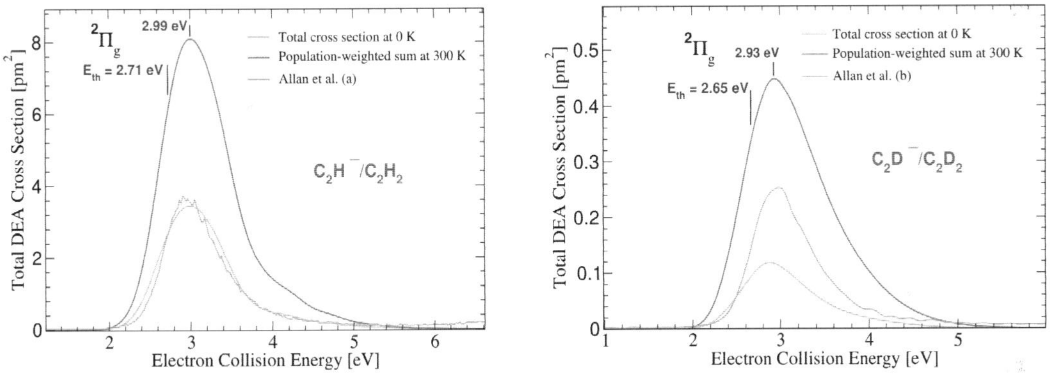

Figure 1. Total cross sections for DA to C_2H_2 (left) and C_2D_2 (right) at 300 K and 0 K compared to experiment [12].

More recently, we have performed nuclear dynamics calculations on C_2H_2 and C_2D_2 to study the isotope effect in DA [11]. Our previous calculations at 0 K led to an isotopic ratio of the cross section for C_2H_2 compared to the cross section for C_2D_2 of 28.9, a factor of 2 higher than recent measurements [12] . Since this reaction proceeds by bending, and this mode is populated at room temperature (the temperature where the experiments were performed), the discrepancy was attributed to the contribution of higher vibrational modes. We included the four lowest bending vibrational states that had non-vanishing populations at the experiment temperature of T = 300 K. The resulting ratio is found to be 17.9 in closer agreement to the experimental value. These results are shown in Figure 1. Although the agreement in absolute magnitude is not as good, this is to be expected since these calculations were carried out at the static-exchange level. In general, the addition of polarization and correlation effects causes changes in the resonance parameters. However, although the value of Γ was found to be smaller by a factor of 0.93 at equilibrium geometry when we performed a Relaxed Self-Consistent-Field (RSCF) calculation, there was only a small shift in the resonance postion. We attempted to obtain a rough estimate of the DA cross section by rescaling the Γ surface using the 0.93 factor. This results in a cross section maximum of 4.9 pm^2 around 3 eV for HCCH. Future calculations will address this as well as the Feshbach resonances that are known to occur at higher energies.

Journal of Physics: Conference Series **204** (2010) 012001 doi:10.1088/1742-6596/204/1/012001

3.2. HCN/HNC

We have previously carried out studies on dissociative electron attachment to ClCN and BrCN [13] . These calculation showed less effect of bending compared to the calculations on HCCH, since the Σ resonant states occurred at a lower energy relative to the ion. Even upon bending dissociation proceeded directly on the A′ state associated with the Σ state in linear geometry, with little interaction with the A′ state arising from the Π state when bent. We decided to study HCN and its isomer HNC to see how they compare to ClCN and BrCN. These systems are isoelectronic to HCCH. We were interested in a comparison to this system as well as studying the isomer effect between HCN and HNC. There exist previous electron scattering calculations on these systems [14, 15, 16]. However, we were interested in studying the dynamics of dissociative electron attachment in these systems.

We carried out calculations at the static exchange level for this system in Jacobi coordinates (three dimensions). Note that these systems share a common potential energy surface, that is, for an angle of zero degrees the system is HCN, at 180 degrees it is HNC. Figure 2 shows a two-dimensional slice of the potential energy surfaces as a function of two coordinates, the distance from H to the CN center of mass (R) and the Jacobi angle (θ). The C-N distance (r) is fixed at the equilibrium value of the neutral. As in the case of HCCH, in linear geometry the Π resonant state lies parallel to the neutral, while the Σ resonant state is higher in energy, but dissociative. Upon bending, the Π state splits into A″ which remains parallel to the neutral, and A′ which interacts with the A′ arising from the Σ. This produces a path to dissociation which has a barrier. The barrier reduces in height as the bending angle is increased.

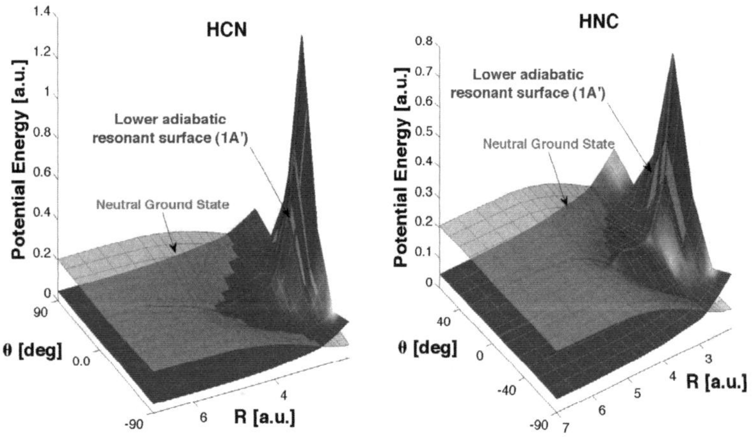

Figure 2. Potential energy surfaces for the neutral and lower adiabatic anion resonant state. HCN (left) and HNC (right). The CN distance has been fixed at the equilibrium distance of the neutral.

However, the dynamics in these systems was quite different than that observed in HCCH, as well as BrCN and ClCN. As in HCCH, the dissociation occurs on the lowest A′ surface. This surface correlates in linear geometry to the Π state at bond distances close to and smaller than equilibrium , but to the Σ state at larger separations. However, instead of bending as was the case in HCCH, the molecule tunnels through the barrier to dissociation. We found that the DA cross section was larger for HCN than HNC. This can be explained by the barrier to dissociation which is lower in energy and more narrow as a function of internuclear separation in the HCN case.

3.3. HCCCN

Experiments on dissociative electron attachment (DA) to HCCCN below 12 eV have led predominantly to formation of $CCCN^-$, CN^-, HCC^- and CC^- negative ions [17]. It has been concluded that these fragments result mainly from the decay of a π^* shape resonance state upon electron attachment that involves distortion of the symmetry of the linear neutral molecule. In order to study the dynamics of dissociation in these channels, we subdivided the molecule into three fragments (H), (CC) and (CN); therefore, four internal coordinates consisting in the distances between the center of masses of (H) and (CC) fragments, (CC) and (CN) fragments, the (H)-(CC) angle and the (CC)-(CN) angle are included in the calculation. We have performed electron scattering calculations at the static exchange level using the complex Kohn variational method [3] to determine the resonance energies and autoionization widths for various geometries of the system and construct the complex potential energy surfaces relevant to the metastable $HCCCN^-$ ion. In Figure 3, the eigenphase sum for elastic scattering of electrons with energy

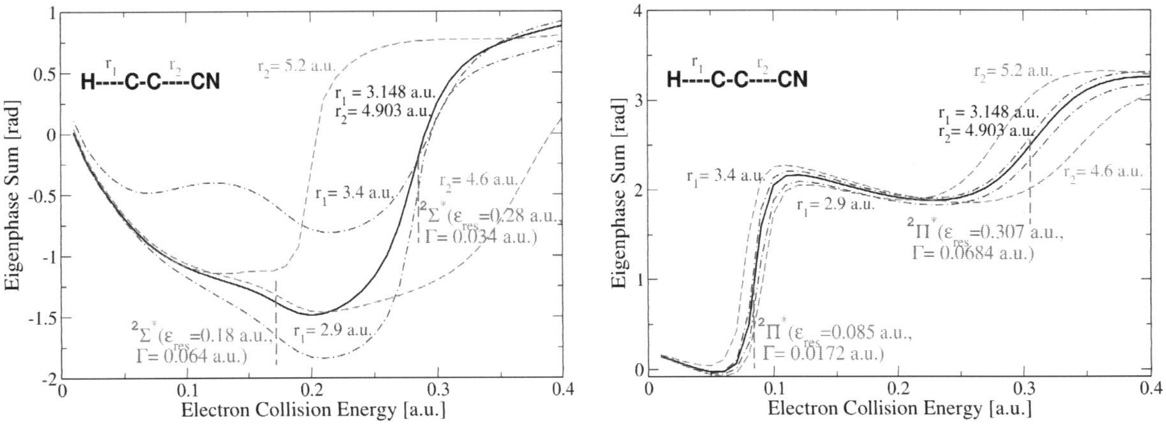

Figure 3. Eigenphase sum for elastic scattering of electrons as a function of incident electron scattering energy, solid curve equilibrium geometry, dashed curves other geometries, left figure Σ symmetry, right figure Π symmetry

below 10 eV are shown. These calculations reveal two σ^* and two π^* resonances at energies consistent with ion fragments count peaks observed experimentally. As can be seen in the figure there are qualitative similarities in the topology of the adiabatic resonant surfaces of HCCCN and that of HCCH and HCN/HNC where the σ^* and the π^* states cross in linear geometry. The eigenphase sum for Σ symmetry show that these resonant states are dissociative, while the Π are not. It is expected that upon bending of the molecule the π^* resonant surfaces will split into an A' and A'' manifolds forming a conical intersection at the crossing point. Further calculations are needed to include the angular degrees of freedom so as to describe the four main DA channels observed experimentally and to correctly describe the dissociation dynamics. Once the surfaces are constructed it will be possible to compute the dissociation dynamics on the coupled resonant surfaces and obtain the DA cross sections.

Journal of Physics: Conference Series **204** (2010) 012001 doi:10.1088/1742-6596/204/1/012001

4. Examples: Dissociative Recombination

4.1. H_2O

The dissociation dynamics in recombination of H_2O^+ with low-energy electrons has been the subject of a number of experiments, especially in the case of 'zero' energy electrons. In the case of an ion storage ring, 'zero' energy is defined as the lowest collision energy that can be obtained in the ion ring. This varies between rings, but is roughly on the order of a few meV [2]. The X $^1A_1$ ground state of H_2O has the dominant configuration $(1a_1)^2(2a_1)^2(1b_2)^2(3a_1)^2(1b_1)^2$. The electron in the highest occupied molecular orbital (HOMO) is bound by 12.6 eV, and removal of this leads to the formation of H_2O^+ in its $^2B_1$ ground state. Dissociative recombination with a zero eV electron gives:

$$\begin{aligned} H_2O^+ + e^- \quad &\rightarrow OH + H(^2S) + 7.5\text{ eV} \\ &\rightarrow O(^3P) + H_2 + 7.6\text{ eV} \\ &\rightarrow O(^3P) + H(^2S) + H(^2S) + 3.04\text{ eV} \\ &\rightarrow O(^1D) + H(^2S) + H(^2S) + 1.07\text{ eV} \end{aligned}$$

where the energies are given for OH and H_2 in their zeroth vibrational levels and with negligible rotational energy. The ion storage ring experiments [18, 19] indicate a strong preference for three-body breakup, while the branching ratios from flowing afterglow indicate the two-body breakup dominates [20]. In addition, DR of HDO^+ shows a 2:1 preference to break the OH rather than the OD bond. There also exists data on the energy and angular distributions of the fragments in the case of three-body breakup at 'zero' energy. There is however no theory available for this system, no experimental data available at other energies, and limited information as to the branching between the final states of the O atom fragment. The existence of the three-body channel indicates that one must go beyond a simple one-dimensional picture in this system. The limited angular fragmentation pattern [19] also suggests it will be necessary to use a full three-dimensional surface to understand the dynamics of this reaction.

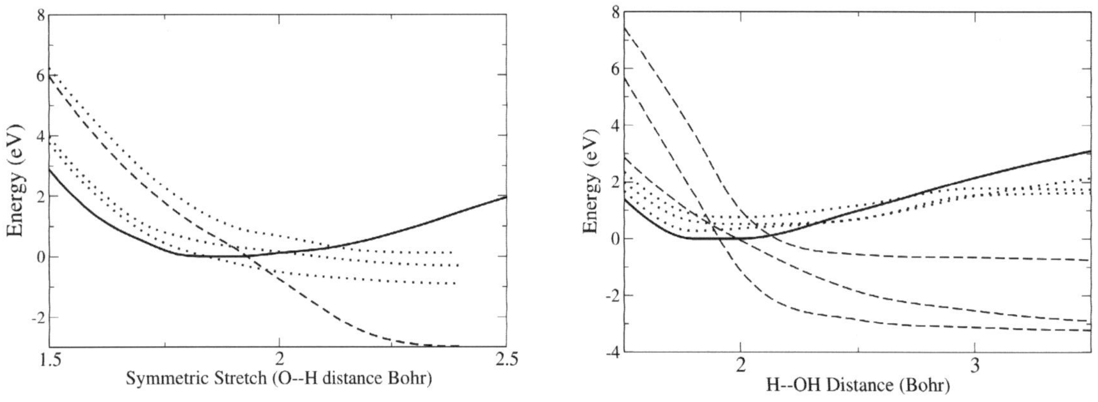

Figure 4. Potential energy curves for resonant states of H_2O compared to the ground state of the ion as a function of (left) symmetric stretch and (right) asymmetric stretch. The solid curve is the ground state of the ion, the dotted curves and dashed curves are resonances that are members of two separate Rydberg series that converge to different states of the ion. The curves have been diabatized.

We have carried out preliminary fixed nuclei Complex Kohn calculations on water to obtain the potential energy surfaces for the low-lying resonant states. The results for these calculations are shown in Figure 4 where the resonant states are shown compared to the ion. These calculations were carried out as a function of symmetric stretch (shown on the left side of the figure), where the bond angle is held fixed at the equilibrium values and both O–H bond distances are changed together. Also we show results as a function of asymmetric stretch, where again the bond angle is fixed, but now one O–H bond distance is fixed and the other O–H is varied. These calculations show the existence of a number of resonant states that exhibit favorable crossings with the ion. In both symmetric and asymmetric stretch, there exist resonances that are members of two separate Rydberg series that converge to different excited states of the ion, and experience a strong interaction near the Franck Condon region.The curves shown have been diabatized by following the leading term in the configuration-interaction wave function to show the interaction. Further calculations are needed to form the three-dimensional potential energy surfaces and to compute the dynamics.

5. Summary

As we have shown in these examples, electron resonances are ubiquitous in low-energy scattering from molecules and molecular ions and provide an efficient path for channeling energy into nuclear motion In order to treat this process accurately, *ab initio* theory requires both an accurate treatment of electronic scattering and nuclear dynamics. This is now possible for diatomic and small polyatomic systems. We have also show that in order to understand how energy flows in a polyatomic systems it is critical to go beyond one-dimensional models. Much work is still needed, both experimentally and theoretically to truly understand these critical collision processes.

Acknowledgments

Å. L. acknowledges support from The Swedish Research Council and A.E.O acknowledges support from the National Science Foundation under Grant No.PHY-05-55401 and the U. S. DOE Office of Basic Energy Science, Division of Chemical Science.

References

[1] *Plasma Processing of Materials:Scientific Opportunity and Technological Challenges* National Research Council (National Academy Press, Washington, DC, 1991); *Database Needs for Modeling and Simulation of Plasma Processing" National Research Council* (National Academy Press, Washington, DC,1996)
[2] For a recent review, see *Dissociative Recombination of Molecular Ions* 2008, Larsson M and Orel A E Cambridge University Press
[3] Rescigno T N, Lengsfield B H and McCurdy C W 1995 in *Modern Electronic Structure Theory*, edited by D. R. Yarkony (World Scientific, Singapore,1995), Vol. **1**, p. 501
[4] O'Malley T F 1966 *Phys. Rev.* **150** 14
[5] Hazi A U, Rescigno T N, and Kurilla M 1981 *Phys. Rev.A* **23** 1089
[6] McCurdy C W and Turner J L 1983 *J. Chem. Phys.* **78** 6773
[7] Beck M H, Jäckle A, Worth G A and Meyer H-D 2000 *Physics Reports* **324** 1
[8] May O, Fedor J, Ibanescu B C and Allan M 2008 *Phys. Rev. A* **77**, 040701
[9] Rescigno T N, Trevisan C S and Orel A E 2006 *Phys. Rev. Lett.* **96** 213201
[10] Chourou S T and Orel A E 2008 *Phys. Rev. A* **77** 042709
[11] Chourou S T and Orel A E 2009 *Phys. Rev. A* **80** 034701
[12] May O, Fedor J and Allan M 2009 *Phys. Rev. A* **80** 012706
[13] Royal J and Orel A E 2006 *J. Chem. Phys,* **125** 214307
[14] Jain A and Norcross D W 1985 *Phys. Rev. A* **32** 134
[15] Jain A and Norcross D W 1986, *J. Chem. Phys.* **84** 739
[16] Varambhia H Nand Tennyson J 2007 *J. Phys. B: At. Mol. Opt. Phys.* **40** 1211
[17] Graupner K, Merrigan T L, Field T A, Youngs T G A and Marr P C 2006 *New J. Phys.* **8** 117
[18] Jensen M J, Bilodeau R C, Heber O, Pedersen H B, Safvan C P, Urbain X, Zajfman D and Andersen L H 1999 *Phys. Rev.A* **60** 2970
[19] Rosen S, Derkatch A, Semaniak J, Neau A, Al-Khalili A,Le Padellec A, Vikor L,Thomas R, Danared H, Ugglas M af and Larsson M 2000 *Faraday Discussions* **115**, 295
[20] Rowe B R, Valle F, Queffelec J L, Gomet J C, and Morlais M 1987 *J. Chem. Phys.* **88** 845

Journal of Physics: Conference Series **204** (2010) 012002 doi:10.1088/1742-6596/204/1/012002

The vibrational excitation of hot molecules by low energy electron impact

H. Kato[1], M. Ohkawa[1], M. Hoshino[1], L. Campbell[2], M.J. Brunger[2] and H. Tanaka[1]

[1] Department of Physics, Sophia University, Chiyoda-ku, Tokyo 102-8554, Japan
[2] ARC Centre for Antimatter-Matter Studies, SoCPES, Flinders University, GPO Box 2100, Adelaide, SA 5001, Australia

E-mail: katou-h@hoffman.cc.sophia.ac.jp

Abstract. We report vibrational excitation functions and angular distributions for electron scattering from the ground vibrational quantum (000), the bending vibrational quantum (010) and the unresolved first bending overtone (020) and symmetric stretch (100) modes of the ground-electronic state in hot (750 K) carbon dioxide (CO_2) molecules. The excitation function measurements were carried out at incident electron energies in the range of 1–9 eV, and at the electron scattering angles of 30°, 60°, 90° and 120°.

1. Introduction

Experimental studies on the excitation processes of molecules by low energy electron impact have usually been performed under conditions near room temperature. However, it is very rare that the system or the phenomenon under study e.g. in nature or in a laboratory, corresponds to such an ideal system. In low temperature plasmas (400-1000 K), it has been pointed out that vibrationally excited (hot) molecules play important roles in the behaviour of that plasma. While there have been several investigations, both experimental and theoretical [1] (and references therein), into electron impact excitation of room temperature CO_2, where the predominant scattering occurs from the ground vibrational quantum (000) of the ground electronic state ($^1\Sigma_g^+$), studies into electron scattering from hot CO_2 are still quite limited. We believe that the first quantitative experiment to study low-energy electron scattering from vibrationally excited CO_2 molecules was by Buckman et al. [2], in which total cross sections (TCSs) as a function of temperature were determined using a linear attenuation time-of-flight (TOF) spectrometer. In that work Buckman et al. observed a substantial increase in the TCS at electron energies below 2 eV, which they attributed to enhanced scattering due to the electric dipole moment of the bending (010) mode. No significant change in the TCS was observed at higher energies. Ferch et al. [3] repeated the experiment of Buckman et al., also using a TOF spectrometer, and confirmed the earlier result of direct dipole scattering but also found, in contrast, a pronounced change (increase) in the TCSs at energies from about 3 to 5 eV, where the $^2\Pi_u$ resonant state of CO_2 is formed. The resonance contribution from the vibrationally excited (hot) molecules, again mainly in the (010) mode, was also observed to shift to lower energies, compared to the ground-state case, by about 0.3 eV. This point is important and is used later in our analysis. More recently, Strakeljahn et al. [4] confirmed the results from Ferch et al. With respect to angular distribution measurements for electron scattering from hot CO_2, we are aware of only the data from Johnstone et al. [5], at an impact energy of 4 eV and scattering angle of 30°, and the somewhat more extensive results from Johnstone et al. [6], now at 3.8 eV, but for the scattered electron angles 20°, 30°, 40°, 60° and 80°. These latter results obtained the first inelastic cross sections for scattering from the (010)*→ (020)(100) modes and for superelastic scattering of the (010)* mode to the ground-state (000) level. Note that the asterisk here and throughout this paper denotes scattering from an excited vibrational mode of the CO_2 molecule. Nonetheless the available data in the literature remains quite limited. Indeed part of the rationale for the present study was to try to significantly extend the superelastic and inelastic vibrational cross sections that are available. This is not only important for providing benchmark data against which sophisticated theoretical calculations can be compared, but is also vital when trying to study the role of electron-driven processes on Venus and Mars where CO_2 is one of the two dominant atmospheric species [7]. In recent years, the vibrational excitation functions for the inelastic and superelastic transitions of hot CO_2 were reported by our group [8] for the first time. In this paper, angular distributions of vibrational DCSs in hot CO_2 are now reported.

2. Experimental and analysis details

The experimental apparatus consists of a typical crossed-beam method in conjunction with the relative flow technique [9]. The present experiments were performed with an overall energy resolution of about 30 meV, and with an incident current of the order of a few nA in the impact energy range from 1 eV to 9 eV. The absolute scale is set using the relative flow technique with helium and its well-established cross sections [10] as the standard gas. Note that the relative flow method is corrected for the effect of the target gas temperature on the gas flux by using $J \propto P/\sqrt{T}$, where J is the gas flux, P the target gas pressure and T the temperature of the target gas. The impact energy scale was calibrated against the He resonance at 19.367 eV, as well as the $^2\Pi_g$ resonance of N_2 (v=1) at 0.97 eV [11]. We believe this is accurate to ±10 meV.

We employed a gas resistive heating method (see Figure 1) to produce hot CO_2. The CO_2 gas is heated using a cylindrical cell made of solid Cu, which incorporates a 4 mm long nozzle of 0.6 mm diameter. The cell itself has an internal diameter of 10 mm and a length of 12 mm, with a quartz fibre. The cell is heated by wrapping a resistive coaxial sheath wire around it, and can reach temperatures up to about 850 K. A magnetic shield, placed around the cell, prevents the field generated by the heater, around which there is wrapped a pipe with flowing water that provides an outer cooling jacket, from penetrating into the interaction region. A thermocouple was set in to a small hole on the cell, and can serve to monitor the cell temperature. From an earlier experiment [12] that produced a gas phase C_{60} beam, it has been confirmed that the gas temperature of the cell can be estimated approximately to be the same as that for the CO_2 beam at the collision centre (~1.5–2 mm from the cell nozzle exit).

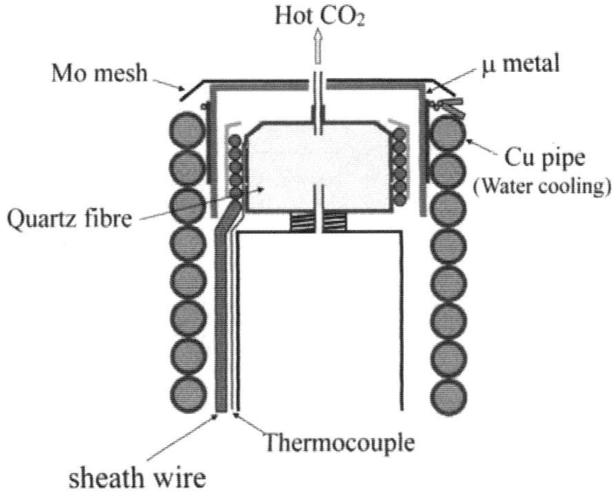

Figure 1. Schematic diagram of our hot CO_2 source.

While sweeping the electron impact energies over the range 1–9 eV, the excitation function experiments were carried out at the respective energy loss values of 0.083 eV and -0.083 eV and at the scattering angles of 30°, 60°, 90° and 120°. Furthermore they were measured at the temperatures of 300, 550 and 750 K. Our experimental results therefore include cross sections from a variety of initial states whose populations depend on the Boltzmann distribution (see Table 1). The result of this process is thus, at each θ, an unresolved absolute excitation function for the (010)*→(000) and (020)*(100)*→(010)* superelastic transitions, and an unresolved absolute excitation function for the (000)→(010) and (010)*→(020)(100) inelastic transitions. We attempt to uniquely deconvolve their respective contributions by using the following assumptions in conjunction with a least squares fit. In the deconvolution, in all relevant cases, we assume that the Principle of Detailed Balance [13] holds. This assumption is expected to be valid, following the work of Johnstone et al. [6]. In the case of the bending mode this means that the condition:

$$E_0 g_0 \sigma_{01}(E_0, \theta) = (E_0 - \Delta E_1) g_1 \sigma_{10}(E_0 - \Delta E_1, \theta) \quad (1)$$

must be maintained in the deconvolution process. Note that in Eq. (1) above g_v is the degeneracy of the mode, ΔE_v is the energy of the vibrational quantum excited, σ_{01} is shorthand notation for the inelastic cross section for the (000)→(010) transition, while σ_{10} denotes the superelastic cross section for the (010)*→(000) transition.

In Figure 2 we therefore show the absolute excitation function results from this process, as measured at θ=90°, for the (000)→(010), (010)*→(020)(100), (010)*→(000) and (020)*(100)*→(010)* transitions. The overall errors on these excitation functions are in the range 20–70 % and include contributions from the statistical accuracy of the data, an uncertainty in our experimental calibration procedures including our normalisation to set the absolute scale, and an uncertainty reflecting the uniqueness of our least squares fit deconvolutions.

Table 1. Population fractions of the vibrational modes in the initial state of CO_2, determined from a standard Boltzmann distribution [14].

Vibrational mode	Energy (meV)	g_v	Population fractions (%)		
			300K	550K	750K
(000)	0.00	1	91.8632	66.4394	49.8724
(01^10)	82.75	2	7.4819	23.1840	27.7222
(02^00)	159.37	1	0.1931	2.3018	4.2357
(02^20)	165.54	2	0.3042	4.0416	7.7001
(10^00)	172.11	1	0.1180	1.7592	3.4779
(03^10)	239.59	2	0.0173	0.8472	2.4485
(03^30)	248.38	2	0.0123	0.7038	2.1371
(11^10)	257.50	2	0.0087	0.5806	1.8558
(00^01)	291.26	1	0.0012	0.1424	0.5504

3. Results and discussion

It is clear from Figure 2 that the $^2\Pi_u$ resonance plays a significant role in enhancing the measured cross sections for all the excitation functions measured as a part of this study. It is also clear that there are no detailed theoretical calculations against which we can compare the present excitation functions. The resonance energy of (010)*→(020)(100) is shifted to about 0.4 eV on the lower energy side compared to (000)→(010), and the cross sections are decreased in magnitude to about half. The resonance energy of (020)*(100)*→(010)* is also shifted to the lower energy side compared to (010)*→(000). In the comparison of the transitions (010)*→(020)(100) and (010)*→(000), from the same initial state, shown by solid line in Figure 2, we find they have similar cross sections, and their peak resonance energies are now only about 0.2 eV different. We note that for the (010) mode the molecule undergoes a 5 degree deformation from its equilibrium geometry. However, even this small deformation can account for the 0.2 eV energy shift noted above. This is explained by the Renner-Teller effect as follows: the CO_2^- state is split to the bent $^2A_1$ and linear $^2B_1$ states with the shift in the resonance energies being dependent on this effect. At the scattering angles of 30°, 60°, 90° and 120° (not shown), we observed that the peak resonance energies all had the same shift as that at the scattering angle of 90°. This observation is in agreement with the resonance energy shift of the TCS results of Ferch et al. [3] and Strakeljahn et al. [4]. Moreover, in the TCS, cross sections from the vibrationally-excited state are about 1.5 times larger than those from the ground vibrational state. On the other hand, in our DCS results, cross sections for vibrational excitation from vibrationally-excited states are not larger than those from the ground vibrational state. Therefore, we suggest that the increased magnitude in the TCS from excited CO_2 is more dependent on the elastic process. This is consistent with results for elastic scattering from excited CO_2 from Johnstone et al. [6]

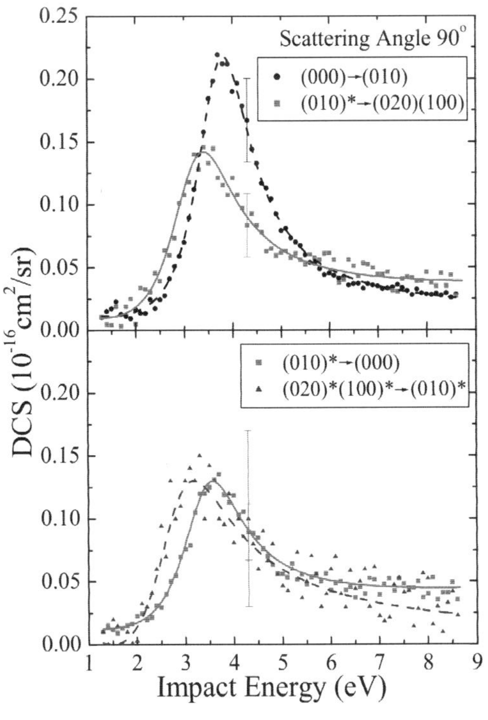

Figure 2. Vibrational excitation functions for the (000)→(010) and (010)*→(020)(100) inelastic transitions and (010)*→(000) and (020)*(100)*→(010)* superelastic transitions of CO_2 at impact energies of 1–9 eV and at a scattering angle of 90°. In each case the line simply serves to highlight the observed energy dependence.

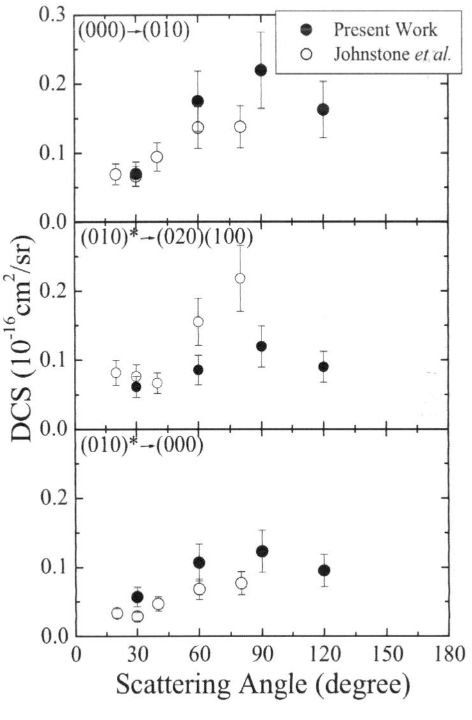

Figure 3. Angular distributions of vibrational excitation DCS for the inelastic and superelastic transitions of CO_2, at an impact energy of 3.8 eV and at scattering angles in the range 30°–120°.

The angular distributions of vibrational excitation DCSs for the inelastic and superelastic transitions at the impact energy of 3.8 eV, from excitation functions, is shown in Figure 3. At all scattering angles of the present work, cross sections for (010)*→(020)(100) and (010)*→(000) are at about the same magnitude and are about half of those cross sections for (000)→(010). The comparison between the present work and Johnstone et al.'s data showed that the cross sections of (000)→(010) and (010)*→(000) are in good agreement within error. But, on the (010)*→(020)(100) transition, the cross sections of Johnstone et al.'s data are larger than the present results at 60° and 90°. A general observation from these results is that cross sections at the impact energy of the resonance energy region show strong enhancement at around 90°. This suggests d-wave like scattering is important here in the reaction mechanism description [15].

4. Conclusions

We have reported absolute excitation functions and angular distributions for the (000)→(010) and (010)*→(020)(100) inelastic and (010)*→(000) and (020)*(100)*→(010)* superelastic transitions for electron scattering from hot CO_2. We hope that the present measurements will now stimulate theorists to investigate the processes we have studied in more detail. Finally we believe that the branching ratio between the resonant vibrational excitation cross sections for ground-state and hot CO_2 scattering, which are the most effective processes for the thermalisation of lower energy electrons, provides important information for the modelling of upper atmospheric phenomena.

Acknowledgements

This work was conducted at Sophia University under the auspices of the Japanese Ministry of Education, Sport, Culture and Technology. H.K. acknowledges the Japan Society for the Promotion of Science (JSPS) for his fellowship as a grant-in-aid for scientific research. L.C. and M.J.B. thank the ARC Centre for Antimatter-Matter Studies for some additional financial support. We all thank the International Atomic Energy Agency (IAEA) for its support of this work as a part of a data base project.

References

[1] Itikawa Y 2002 *J. Phys. Chem. Ref. Data* **31**,749
[2] Buckman S J, Elford M T and Newman D S 1987 *J. Phys.* B **20** 5175
[3] Ferch J, Masche C, Raith W and Wiemann L 1989 *Phys. Rev.* A **40** 5407
[4] Strakeljahn G, Ferch J and Raith W 1995 *Proc. 19th ICPEAC* (British Columbia University Press) p.428
[5] Johnstone W M, Mason N J and Newell W R 1993 *J. Phys.* B **26** L147
[6] Johnstone W M, Brunger M J and Newell W R 1999 *J. Phys.* B **32** 5779
[7] Campbell L, Brunger M J and Rescigno T N 2008 *J. Geophys. Res.* **113** E08008
[8] Kato H, Kawahara H, Hoshino M, Tanaka H, Campbell L, Brunger M J 2008 *Chem. Phys. Lett.* **465**, 31
[9] Tanaka H, Boesten L, Matsunaga D, Kudo T 1988 *J. Phys.* B **21** 1255
[10] Boesten L and Tanaka H 1992 *Atom Data Nucl. Data Tables* **52** 2
[11] Sun W, Morrison M A, Isaacs W A, Trail W K, Alle D T, Gulley R J, Brennan M J and Buckman S J 1995 *Phys. Rev.* A **52** 1229
[12] Tanaka H, Boesten L, Onda K and Ohashi O 1994 *J. Phys. Soc. Jpn.* **63** 485
[13] Fowler R H 1936 *Statistical Mechanics: The Theory of the Properties of Matter in Equilibrium* (Cambridge University Press)
[14] Herzberg G 1945 *Molecular Spectra and Molecular Structure. Infra Red and Raman Spectra of Polyatomic Molecules* vol 2 (New York: Van Nostrand)
[15] Brandson B H and Joachain C J 1983 *Physics of Atoms and Molecules* (New York: Longman Group Ltd.)

Journal of Physics: Conference Series **204** (2010) 012003 doi:10.1088/1742-6596/204/1/012003

Electron impact processes in nitrogen rich atmospheres of the outer solar system

P V Johnson[1], J A Young[1], C P Malone[1,2], M A Khakoo[2], X Liu[1], and I Kanik[1]

[1] Jet Propulsion Laboratory, California Institute of Technology, 4800 Oak Grove Drive, Pasadena, CA 91109-8099, USA
[2] Department of Physics, California State University, Fullerton, CA 92834-6866, USA

Email: Paul.V.Johnson@jpl.nasa.gov

Abstract. Electron collisions with N_2 are predominant components of the interaction between outer solar-system atmospheres and solar photons, solar wind, and magnetospheric electrons. Our collaboration between the Jet Propulsion Laboratory (JPL) and the California State University, Fullerton (CSUF), has devoted a significant effort over the last few years to improving the knowledge base of electron-N_2 excitation phenomena. Here, we highlight the lack of Franck-Condon behaviour in excitation of the Rydberg-valence states, using excitation of the $C\,^3\Pi_u$ state as a case in point. Further, we briefly introduce ongoing work measuring vibrationally-resolved emission cross sections for the Lyman-Birge-Hopfield (LBH) band system of the $a\,^1\Pi_g\,(v') \rightarrow X\,^1\Sigma_g^+\,(v'')$ transitions. Specifically, we present preliminary results for the relative (3,0) emission cross section, which calls the currently accepted LBH shape function of Ajello and Shemansky (1985 *J. Geophys. Res.-Space* **90** 9845) into question.

1. Introduction

It is currently an exciting time to study electron-molecule collisions, with countless opportunities to apply the results towards atmospheric processes in the outer solar system. The Galileo and Cassini missions have provided a wealth of atmospheric data, which continues to be digested. Meanwhile, the Cassini spacecraft remains in orbit around Saturn, providing additional data as it fulfils its extended *Equinox* mission. Beyond Cassini, the Juno mission to Jupiter is expected to launch in August 2011, which will be followed by the recently selected Europa Jupiter System Mission (EJSM) as the next NASA flagship. Juno will carry an Ultraviolet Imaging Spectrograph (UVS) as part of its "Polar Magnetosphere Suite" to study the Jovian aurora (http://juno.wisc.edu/index.html). The EJSM is envisioned as consisting of a NASA-led Jupiter Europa Orbiter (JEO) and Jupiter Ganymede Orbiter (JGO) led by the European Space Agency (ESA). Both anticipate UVSs as part of their science instrument payloads (http://opfm.jpl.nasa.gov). As such, studies of electron-molecule collisions and there subsequent emission (particularly in the UV) will continue to be of great interest to the planetary science and aeronomy communities, as they model and interpret these data.

2. Molecular nitrogen

Molecular nitrogen is the primary atmospheric constituent of Saturn's moon Titan (94%) and Neptune's moon Triton (99.9%). Interaction of these atmospheres with solar photons, the solar wind, and the magnetospheric plasmas of their respective parent bodies, ensures an abundant supply of free electrons in their upper atmospheres. Collisions between these electrons and the atmospheric N_2 result in photon emissions that provide an important diagnostic probe for understanding these environments. Ultraviolet (UV) spectra of Titan and Triton taken by Voyager covered a wavelength range of 50–170 nm. These spectra show strong emission features, which have been interpreted as resulting primarily due to transitions from the $a\,^1\Pi_g$, $b\,^1\Pi_u$, $b'\,^1\Sigma_u^+$, and $c_4'\,^1\Sigma_u^+$ states of N_2 to the $X\,^1\Sigma_g^+$ ground state (i.e., the Lyman-Birge-Hopfield (LBH), Birge-Hopfield I and II, and Caroll-Yoshino bands, respectively) [1]. Recently, De Le Haye *et al* [2] have modelled the heat released into Titan's atmosphere through collisions between photo- and magnetospheric electrons. They show that N_2 electronic excitation processes dominate the net heating rate throughout the upper atmosphere (~800–2000 km). Observations and modelling efforts such as these provide the impetus for laboratory studies of electron-N_2 collision processes.

2.1. Differential and integral excitation cross sections

In the last five years the JPL-CSUF collaboration has conducted an extensive study of electron impact excitation of molecular nitrogen (N_2). Specifically, differential and integral cross sections (DCSs and ICSs, respectively), have been measured using electron energy loss spectroscopy in a crossed beam arrangement for excitation of the $A\,^3\Sigma_u^+$, $B\,^3\Pi_g$, $W\,^3\Delta_u$, $B'\,^3\Sigma_u^-$, $a'\,^1\Sigma_u^-$, $a\,^1\Pi_g$, $w\,^1\Delta_u$, $C\,^3\Pi_u$, $E\,^3\Sigma_g^+$, $a''\,^1\Sigma_g^+$, $b\,^1\Pi_u$, $c_3\,^1\Pi_u$, $o_3\,^1\Pi_u$, $b'\,^1\Sigma_u^+$, $c_4'\,^1\Sigma_u^+$, $G\,^3\Pi_u$, and $F\,^3\Pi_u$ electronic states of N_2 [3-8]. In comparison with other measurements of this kind, this body of work differentiates itself in that it did not directly rely on Franck-Condon (FC) factors in the spectral unfolding of the high-lying, Rydberg-valence states, i.e., the $C\,^3\Pi_u$, $b\,^1\Pi_u$, $c_3\,^1\Pi_u$, $o_3\,^1\Pi_u$, $b'\,^1\Sigma_u^+$, $c_4'\,^1\Sigma_u^+$, $G\,^3\Pi_u$, and $F\,^3\Pi_u$ states (see Ref. [5] for details).

The unfolding distinction is important as the spectroscopy of these states is complicated by very strong Rydberg-valence, Rydberg-Rydberg, and valence-valence interactions within the $^1\Pi_u$ and $^1\Sigma_u^+$ manifolds [9-11], and potentially within the $^3\Pi_u$ manifold [12]. Thus, the assumption of invariant FC factors, implicit in the techniques normally used in the spectral decomposition of electron energy-loss spectra, is invalid in this case. Malone *et al* [7, 8] present vibrationally resolved DCSs and ICSs for the $C\,^3\Pi_u$ state. Figure 1 shows relative excitation probabilities (REPs) of the v'=0–3 levels from Malone *et al* [7], Zubek and King [13], and Poparic *et al* [14] (peak values), normalized so the REP of v'=0 is unity to enable a comparison. Likewise, normalized REPs of the v'=1–3 levels are included for the emission based measurements of Zubek [15] from onset to 17.5 eV, and the peak cross-section REPs of Shemansky *et al* [16], Fons *et al* [17], and Tohyama and Nagata [18] as shaded box-regions, which indicate the spread in both the peak's energy location and REP value for the three experiments.

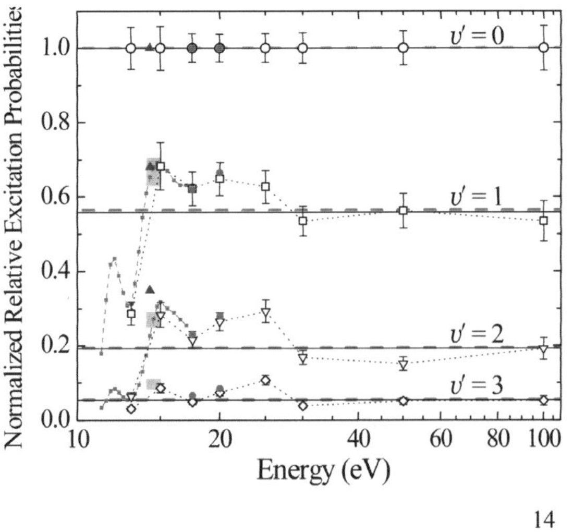

Figure 1. Experimental relative excitation probabilities (REPs) for the $C\,^3\Pi_u$ (v'=0–3) levels normalized so the REP for the v'=0 level is unity. Legend: Malone *et al* [7]–black open symbols with dotted lines; Zubek and King [13]–green circles; Poparic *et al* [14]–blue triangles; Zubek [15]–green squares with dashed lines; emission-based peak REPs [16-18]–grey boxes (see text). The FC factors of Lofthus and Krupenie [19] (black solid lines) and Gilmore *et al* [20] (red dashed lines) are also shown and normalized in a similar manner.

Comparison of our data with Zubek [15] corroborates the steep slope below 15 eV that is partly a threshold effect. The FC factors of Lofthus and Krupenie [19] and Gilmore *et al* [20] are also shown, normalized in a similar manner. The overall accuracy of the present results is substantiated by a comparison with the experimental data of Refs. [13-18]. Minimal disagreement outside of our uncertainty is seen for v'=3, but the v'=0–2 results are in excellent agreement. At impact energies of 50 eV and higher, the experimental REPs agree well with the FC factors of Refs. [19, 20]. However, significant discrepancies are observed between the experimental REPs and the calculated FC factors, explicitly demonstrating the described non-FC factor behaviour.

2.2. Lyman-Birge-Hopfield (LBH) emission cross sections

Previous and ongoing measurements of the Titan system by the Cassini Ultraviolet Imaging Spectrograph (UVIS) requires accurate, absolute $a\,^1\Pi_g$ state excitation ($X\,^1\Sigma_g^+(0) \rightarrow a\,^1\Pi_g(v')$) and LBH band emission ($a\,^1\Pi_g(v') \rightarrow X\,^1\Sigma_g^+(v'')$) cross sections. In the upper atmosphere of Titan, the LBH band system provides a sensitive measurement of N_2 density distribution and photoelectron flux in the dayglow. Therefore, the global energy input to the dayglow can, in principle, be measured from modelling the total LBH intensity.

The excitation cross section for the $a\,^1\Pi_g$ state currently and generally accepted by aeronomers is that of Ajello and Shemansky [21], which was determined based on the measurement of a single relative emission cross section ($a\,^1\Pi_g(3) \rightarrow X\,^1\Sigma_g^+(0)$) [21] as a function of electron impact energy. After obtaining the relative LBH (3,0) emission cross section, it was put on an absolute scale by normalizing the observed N I (119.99 nm) intensity to that of the H_2 Lyman-α cross section in a relative flow experiment at 200 eV. A collision strength model, specifically tailored to better reproduce the shape of their (3,0) measurement [21], was fit to the (3,0) cross section, providing the parameters needed to deduce partial cross sections for other individual transitions as well as for the entire band system via the collision strength model. Given their deduced total LBH emission cross section, a direct excitation cross section was produced by estimating the effects of predissociation and cascade [21]. Apart from potential issues related to the normalization of the Ajello and Shemansky [21] cross section (i.e., the H_2 Lyman-α cross section has since been revised downward by ~14% at 100 eV [22, 23], reliance on a lifetime dependent *glow* model, etc.), recent work suggests the shape of the Ajello and Shemansky (3,0) emission cross section may be inaccurate, thus influencing the modelled shape of the entire LBH band system.

Figure 2 shows three recent measurements of the (3,0) relative emission cross section. These measurements were conducted on two separate apparatus, which are very similar in overall design. In both apparatus, the target molecular species is introduced into a vacuum chamber as a swarm (*static gas*) and allowed to interact with a magnetically collimated, ~0.5 eV full-width at half-maximum (FWHM) electron beam. The resulting photons are collected by a vacuum spectrometer using a

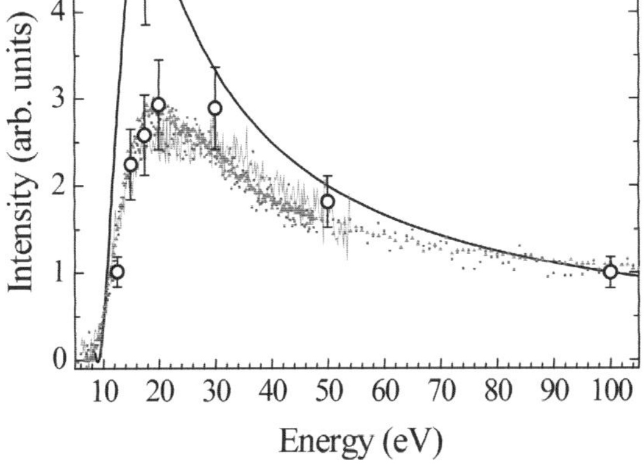

Figure 2. Relative N_2 $a\,^1\Pi_g(3) \rightarrow X\,^1\Sigma_g^+(0)$ emission cross sections (i.e., LBH (3,0)) normalized at 100 eV. Preliminary results were measured in swarm mode on two different instruments (0.2 m and 3 m spectrometer systems) at various pressures. Legend: present data–red triangles (0.2 m; 14 μTorr); present data–blue squares (0.2 m; 2.4 μTorr); present data–green line (3 m, 2.5 μTorr); Ajello and Shemansky [21], modelled cross section–black line; Johnson *et al* [3], ICSs–open circles.

1200 g mm^{-1} grating and detected by a CsI coated channel electron multiplier. A suitably biased Faraday cup is used to measure electron beam current and minimize back-scattered electrons. The main distinction between these instruments is the resolution of the spectrometer used to interrogate the UV emission. Specifically, one instrument is equipped with a 0.2 m focal length spectrometer while the other is equipped with a 3 m spectrometer with ultimate resolutions of ~0.5 nm and ~3.3 pm (FWHM), respectively. Shown in Figure 2 are relative emission cross sections measured on the 0.2 m instrument at 14 μTorr and 2.4 μTorr and on the 3 m system at 2.5 μTorr. Although preliminary, these measurements are very self-consistent in terms of the cross section shape. Further, when compared with the ICSs for excitation of the $a\,^1\Pi_g$ state (full electronic manifold) of our earlier energy loss measurements [3] (arbitrarily normalized to the present data sets at 100 eV to facilitate a shape comparison), we again see excellent shape agreement. However, as seen in the figure, these data are markedly different than the (3,0) shape reported by Ajello and Shemansky [21] regardless of the scaling (here, normalized to the present data sets at 100 eV) used to best match the shape. Further measurements are in process.

As mentioned earlier, only a single vibrational transition was measured by Ajello and Shemansky [21], i.e., the (3,0) emission. Figure 9 in their paper shows the modelled emission cross sections for the (v'=0–6,v''=0) transitions, which were used to construct their total cross section. Comparison of these cross sections shows that, at roughly the maxima of their curves (~17.5 eV), ~80% (or more) of the total reported cross section was inferred by the modelled emission cross sections for the (0–2,0) and (4–6,0) transitions (with the remainder from the measured (3,0) emission). Additionally, Ajello and Shemansky [21] show (their Table 5b) excitation cross sections for the v'=0–14 levels at 100 eV, which also indicate that the v'=3 cross section accounts for only ~17% of the total. Clearly, direct measurement of a greater percentage of the total emission (i.e., more than a single vibrational transition) is crucial for proper and reliable understanding of the *total* LBH emission and excitation cross sections (i.e., sum of all emissions in the band). As discussed in the previous section, even if one ignores effects specific to emission, near threshold effects can significantly alter the energy dependence of individual vibronic transitions. We are currently exploiting the resolution of the 3 m spectrometer system to directly measure excitation functions for various vibrational bands of the LBH system in order to improve the fidelity of our understanding of the LBH system.

Acknowledgement
The work was performed at the Jet Propulsion Laboratory (JPL), California Institute of Technology, under a contract with the National Aeronautics and Space Administration (NASA) and at the California State University, Fullerton (CSUF). Financial support through NASA's Planetary Atmospheres program, NASA's Outer Planets Research program, and the National Science Foundation (Grant No. NSF-PHY-RUI-0653452) is gratefully acknowledged.

References
[1] Strobel D F and Shemansky D E 1982 *J. Geophys. Res.-Space* 87 1361
[2] De La Haye V, Waite J H, Cravens T E, Bougher S W, Robertson I P and Bell J M 2008 *J. Geophys. Res.-Space* 113 A11314
[3] Johnson P V, Malone C P, Kanik I, Tran K and Khakoo M A 2005 *J. Geophys. Res.-Space* 110 A11311
[4] Khakoo M A, Johnson P V, Ozkay I, Yan P, Trajmar S and Kanik I 2005 *Phys. Rev. A* 71 062703
[5] Khakoo M A, Malone C P, Johnson P V, Lewis B R, Laher R, Wang S, Swaminathan V, Nuyujukian D and Kanik I 2008 *Phys. Rev. A* 77 012704
[6] Malone C P, Johnson P V, Kanik I, Ajdari B and Khakoo M A 2009 *Phys. Rev. A* 79 032704
[7] Malone C P, Johnson P V, Kanik I, Ajdari B, Rahman S S, Bata S S, Emigh A and Khakoo M A 2009 *Phys. Rev. A* 79 032705

[8] Malone C P, Johnson P V, Young J A, Liu X, Ajdari B, Khakoo M A and Kanik I 2009 *J. Phys. B.: At. Mol. Opt. Phys.* submitted

[9] Lefebvre-Brion H and Field R W 2004 *The Spectra and Dynamics of Diatomic Molecules* (San Diego: Elsevier)

[10] Spelsberg D and Meyer W 2001 *J. Chem. Phys.* 115 6438

[11] Stahel D, Leoni M and Dressler K 1983 *J. Chem. Phys.* 79 2541

[12] Sprengers J P, Ubachs W and Baldwin K G H 2005 *J. Chem. Phys.* 122 144301

[13] Zubek M and King G C 1994 *J. Phys. B: At. Mol. Opt. Phys.* 27 2613

[14] Poparic G, Vicic M and Belic D S 1999 *Chem. Phys.* 240 283

[15] Zubek M 1994 *J. Phys. B.: At. Mol. Opt. Phys.* 27 573

[16] Shemansky D E, Ajello J M and Kanik I 1995 *Astrophys. J.* 452 472

[17] Fons J T, Schappe R S and Lin C C 1996 53 2239

[18] Tohyama Y and Nagata T 2005 *J. Phys. Soc. Jpn.* 74 326

[19] Lofthus A and Krupenie P H 1977 *J. Phys. Chem. Ref. Data* 6 113

[20] Gilmore F R, Laher R R and Espy P J 1992 *J. Phys. Chem. Ref. Data* 21 1005

[21] Ajello J M and Shemansky D E 1985 *J. Geophys. Res.-Space* 90 9845

[22] Liu X M, Shemansky D E, Ahmed S M, James G K and Ajello J M 1998 *J. Geophys. Res.-Space* 103 26739

[23] van der Burgt P J M, Westerveld W B and Risley J S 1989 *J. Phys. Chem. Ref. Data* 18 1757

Journal of Physics: Conference Series **204** (2010) 012004 doi:10.1088/1742-6596/204/1/012004

Recent progress in the theory of dissociative attachment: From diatomics to biomolecules

Ilya I. Fabrikant

Department of Physics and Astronomy, University of Nebraska-Lincoln, Lincoln, NE 68588, USA

E-mail: `ifabrikant1@unl.edu`

Abstract. We present a summary of recent progress in theoretical studies of low-energy dissociative electron attachment (DEA) to halogen molecules and polyatomic molecules based on the resonance R-matrix theory. It explains many observed features in DEA cross sections including low-energy behavior, threshold resonances and cusps. It also gives correct description of the temperature dependence of the attachment rate coefficients. More recently the theory was applied to two molecules of biological interest, formic acid and glycine. DEA mechanisms in these systems are very similar to those in hydrogen halides.

1. Introduction

Dissociative electron attachment (DEA) processes, in spite of their importance for plasma physics [1, 2, 3], environmental science [4, 5, 6] and radiation damage [7, 8, 9], still do not receive enough attention from theorists. A substantial progress has been made in experimental studies of these processes, mainly due to development of new experimental techniques involving electron beams with high energy resolution [10]. This allowed detailed investigation of threshold phenomena in low-energy dissociative attachment to diatomic and polyatomic molecules: vibrational Feshbach resonances (VFRs) and threshold cusps.

Theoretical description of DEA process requires solving a challenging nuclear dynamics problem involving motion in a nonlocal complex potential [11]. For polyatomic cases a simplifying assumption is always involved: either the nuclear motion is described in the local approximation [12], or a one-mode approximation is used. The local approximation is justified when the resonant state responsible for DEA has a relatively small decay width. The one-mode approximation can be used if there is experimental or theoretical evidence that the DEA process is dominated by one vibrational degree of freedom corresponding to the reaction coordinate. A typical example is a molecule of the type CX_3Y where X stands for the H or F atom and Y for a halogen atom. For a description of the reaction

$$e + CX_3Y \rightarrow CX_3 + Y^- \qquad (1)$$

the theoretical approach based on inclusion of only C–Y symmetric stretch turned out to be successful [13, 14, 15, 16]. It adequately describes the electron energy dependence and the vibrational energy dependence of the DEA cross sections, and also predicts threshold resonances and cusps which are confirmed by experiments with highly energy-resolved electron beams.

The current state of the theory in principle allows a complete *ab initio* solution in diatomic cases. For example, a complex nonlocal dynamics has been completely investigated for hydrogen halides, and a high level of agreement between the theory and experiment has been achieved [17, 18, 19, 20]. In spite of this, some diatomic cases still present a big challenge. Surprisingly, this is the case for relatively simple halogen molecules. The semiempirical R-matrix calculations [21, 22] for DEA to the Cl_2 molecule describe very well the dependence of cross section on electron energy and vibrational temperature. At the same time more recent *ab initio* calculations [23] yield a weaker electron energy dependence and vibrational temperature dependence than that observed in the experiments. The case of low-energy DEA to the even simpler F_2 molecule still presents a big challenge [24]. In spite of a 28-year history of theoretical research on this process, there are still major challenges in bringing theory and experiment together.

The present article summarizes major advances in the theory of DEA processes made by the author and his collaborators since the publication of the review of Hotop *et al.* [10]. Our approach to description of DEA is based on the resonance R-matrix theory [25] which is equivalent to the nonlocal complex potential theory [11]. In this review we discuss only calculations carried out in the one-mode approximation, although some progress on inclusion of several vibrational degrees of freedom has recently been made [26, 27].

2. Attachment to halogen molecules

Until this decade Cl_2 presented an enigma because electron-beam experiments showed an attachment feature typical of zero-energy resonances, though weak, while electron capture into the lowest Cl_2^- electronic state requires a *p*-wave process due to its ${}^2\Sigma_u^+$ symmetry [28]

$$e^- + \mathrm{Cl}_2(X^1\Sigma_g^+) \to \mathrm{Cl}_2^-(X^2\Sigma_u^+) \to \mathrm{Cl}(^2P) + \mathrm{Cl}^-(^1S).$$

This situation was even more unclear for the analogous case of F_2, because the 1987 high-resolution (6 meV) data of Chutjian and Alajajian [29] were demonstrating *s*-wave attachment, since the cross section σ fit the Wigner $\sigma \propto E^{-1/2}$ law appropriate for s-wave attachment as a function of electron energy E.

The problem with Cl_2 was resolved in calculations [21] where it was found that the observed attachment resonance peaked not at zero energy but at 50 meV. The cross section followed the Wigner *p*-wave threshold law, $\sigma \propto E^{1/2}$, in rising from zero energy, but dropped rapidly beyond 50 meV due to an unusual rapid decline in the Franck-Condon factor between anion and neutral with increasing electron energy. The resulting cross section peak predicted to lie at 50 meV could easily be mistaken for a zero-energy resonance in experiments which typically have energy resolutions of 50 meV and higher. Explicit verification of these calculations came from high-resolution (1 meV) experiments by Barsotti *et al.* [30] and Ruf *et al.* [22].

More recently the theoretical results for low-energy attachment to Cl_2 were confirmed by flowing-afterglow Langmuir-probe measurements [31]. The calculated temperature dependence of the attachment rate coefficient is almost in perfect agreement with the experiment, both in terms of the shape and absolute values. It is interesting, however, that at higher temperatures (above about 1000 K) the theoretical values are somewhat lower than the experimental, although this observation is pertinent to only two measured temperature points, and therefore perhaps not very conclusive. A similar discrepancy was earlier observed for attachment to CF_3Br [32], but this was attributed to multimode effects in this target. In any event, one might suggest that the theory does not describe adequately the growth of DEA cross section with the vibrational energy. In this connection it is important to note that in recent *ab initio* calculations [23] of DEA to Cl_2 the obtained dependence of DEA cross sections on vibrational energy is even weaker than in the R-matrix calculations [21, 22], and rotational effects were found to be not important [23]. Therefore, although the Cl_2 problem is basically solved, for complete clarification further *ab initio* calculations are necessary.

Experiments from the Kaiserslautern group [24] have recently provided high-resolution (5-meV) data for F_2 which clearly show a *p*-wave threshold behavior for attachment through the F_2^- (Σ_u^+) ground state, with a peak at 31 meV, in contrast with older work [29]. However, the R-matrix theory that was so successful for Cl_2 failed to properly describe attachment to F_2. R-matrix calculations with available F_2^- potential energy curves result in a peak position at about 170 meV, consistent with previous dissociative attachment calculations [33, 34, 35]. Attempts to modify the anion curves to make the peak position consistent with the measurements [24] result in unrealistic potential curves and very small attachment rate constants that are inconsistent with drift tube measurements [36].

These studies are illustrated in Figs. 1 and 2 where we present potential energy curves as functions of internuclear distance R given relative to the equilibrium internuclear separation R_e, and corresponding DEA cross sections. Calculations with the anion curve similar to that obtained by *ab initio* calculations [35] produce the cross section which peaks at $E = 170$ meV, similar to the result of Brems *et al.* [35]. The cross section which peaks at 50 meV with a significant absolute value can be obtained from the curve which follows very closely the neutral curve to the left of the crossing point, and does not give a deep enough minimum consistent with *ab initio* calculations [37]. Attempts to shift the peak further to lower energies lead to even more unrealistic potential energy curves and very small DEA cross sections.

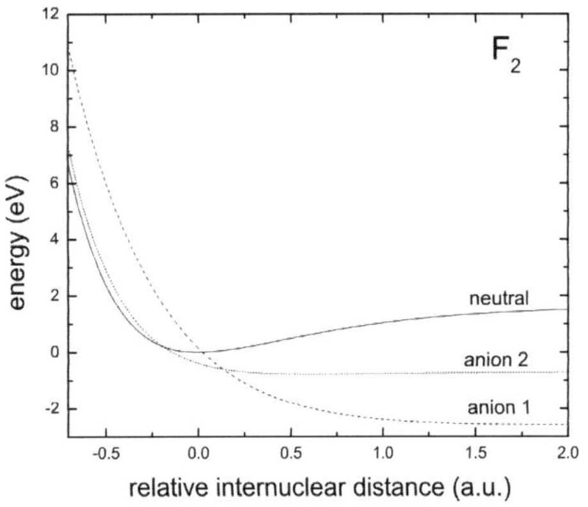

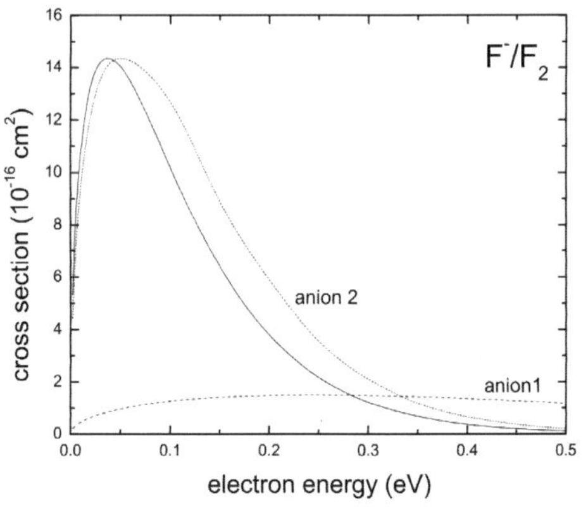

Figure 1. Potential energy curves for F_2. The curve anion 1 corresponds to *ab initio* calculations [35], the curve anion 2 to calculations giving a peak at 50 meV.

Figure 2. DEA cross sections corresponding to the anion curves shown in Fig. 1. Solid line, a fit to measured yield [24] up to $E = 0.1$ eV normalized to the theoretical peak value .

More accurate *ab initio* calculations on this problem are certainly warranted. In particular it is possible that the R-dependence of the anion curve is more complicated than that represented by a simple Morse-potential parametrization used in the semiempirical R-matrix calculations.

3. Attachment to polyatomic molecules

The one-pole R-matrix theory [25] is very well suited for studying resonances and threshold effects in low-energy DEA to polyatomic molecules. However, even in the one-mode approximation, a challenging part in this approach is the calculation of the R-matrix surface amplitude generating the resonance width as described in the early review [38] of the R-matrix theory. In most calculations performed so far the surface amplitude was adjusted empirically in order to reproduce observed vibrational excitation cross sections or thermal attachment rate coefficients. After this parameter is fixed, structures in DEA and vibrational excitation can be calculated with high accuracies using the quasiclassical version [25, 39] of the R-matrix theory. The threshold structures are determined by the long-range electron-molecule interaction which in R-matrix theory is included in the logarithmic derivative of the electron wave function in different vibrational channels.

3.1. Summary of recent results for CX_3Y molecules

Recent studies [32, 40, 41, 42] of the processes of the type (1) are based on adjusting the R-matrix surface amplitudes to measured thermal attachment rate coefficients. Results obtained for DEA to CF_3I, CF_3Br and CH_3Br illustrate some typical features observed in the process (1).

- Low-energy DEA is dominated by formation of the halogen ion Y^-. Because of the high affinity of the halogen atom, all three processes are exothermic, and the DEA cross section at low energy behaves as E^{-x} where $x > 0.5$ is the threshold exponent whose value depends on the dipole moment of the molecule [43]. At ultralow energies below few meV the threshold exponent turns to $x = 0.5$ corresponding to the Bethe-Wigner threshold law. This happens because of molecular rotations – the electron-molecule interaction time becomes so long that molecular dipole moment is averaged out and does not influence the attachment process.
- The vibrational energy dependence of the DEA cross section is determined by the position of the curve crossing between the neutral and anion potential energy curves. In CF_3I the crossing point is very close to the equilibrium internuclear separation, and the vibrational energy dependence of the DEA cross section is very weak. The effect becomes very significant in case of CF_3Br and really dramatic for CH_3Br. In case of CF_3Br the anion curve crosses the neutral at the right turning point for the $\nu_3 = 3$ state [32]. This is a typical case for a strong temperature dependence of the DEA cross section. In case of CH3Br the curve crossing occurs at even higher internuclear separation corresponding to the right turning point for the $\nu_3 = 5$ state. Here we expect very low cross sections for attachment to the ground state (or for attachment at low target temperatures), and the temperature effect should be even stronger than for CF_3Br.
- The potential curve crossing situation for CF_3Br and CH_3Br is also typical for formation of the vibrational Feshbach resonance (VFR) [10] which happens when the electron is captured by the long-range (dipolar plus polarization) field of the molecule with its simultaneous vibrational excitation. However, the dipole moment of CF_3Br too small for VFR formation. Because of the substantial dipole moment of CH_3Br, VFRs appear in this case, as was predicted theoretically [16] and confirmed recently by measurements [24]. In two other molecules, CF_3I and CF_3Br, cusps are observed at vibrational excitation thresholds.
- Threshold structures (cusps and VFRs) are also seen in vibrational excitation and differential elastic scattering cross section [32, 40, 41, 42].

3.2. Asymmetric dissociation

So far we have been discussing DEA processes dominated by one, obviously dominating, vibrational coordinate. However, there are cases when the choice of the dominating coordinate is not obvious. In particular it might happen that the DEA process goes through formation of a totally symmetric resonant anion state, but this resonance produces an asymmetric dissociation. For example, DEA to the chloroform molecule at low energies is dominated by a shape resonance of A_1 symmetry [44, 45]. Therefore one should expect that the resonance drives only totally symmetric vibrations: C-H stretch, symmetric C-Cl stretch, and symmetric CCl_3 deformation. However, dissociation of Cl^- is an asymmetric process, therefore for a rigorous calculation of the

DEA process a multimode treatment, including the interaction between the vibrational modes, is necessary.

Still, with the goal to describe the correct threshold behavior of low-energy DEA to $CHCl_3$, we adopted a simplified approach [46]. First we identify the symmetric vibrational mode which is most important for the DEA process. High-resolution experimental data [46] obtained with the Laser Photoelectron Attachment method indicate a strong coupling between the attachment channel and the symmetric CCl_3 deformation mode ν_3. This coupling appears in the form of a pronounced Wigner cusp in the measured DEA cross section at the threshold for the ν_3 excitation.

Based on this observation and the observation of a direct anion dissociation process following Rydberg electron transfer to $CHCl_3$ [47], we assume that the electron capture initially drives the ν_3 mode and that the primary temporary anion state is stabilized rapidly by direct dissociation along the ν_3 coordinate. Then, mediated by intramolecular vibrational energy redistribution (IVR), the excess energy is channelled into the C-Cl stretch which eventually leads to the dissociation into the fragments Cl^- and $CHCl_2$. The attachment process is thus treated in a one-dimensional approximation, assuming coupling of the anion state with only the ν_3 mode. This approach is similar to our effective-range-theory treatment of electron attachment to SF_6 [48].

The theoretical results for different gas temperatures obtained from this model are presented in Fig. 3, together with experimental data obtained at room temperature by combined Laser Photoelectron Attachment - Extended Laser Photoelectron Attachment (LPA/EXLPA) method [49]. The R-matrix surface amplitude was adjusted to reproduce the measured [46] attachment rate coefficient, $k(T = 300\text{K}) = 3.9(2) \times 10^{-9}$ cm^3/s. Because of a very strong variation of the DEA cross section with the vibrational quatum number ν_3 [46], we observe a strong temperature dependence in the low-energy region. However, the position of the peak between 0.4 and 0.5 eV virtually does not change with the gas temperature that is confirmed by measurements [46].

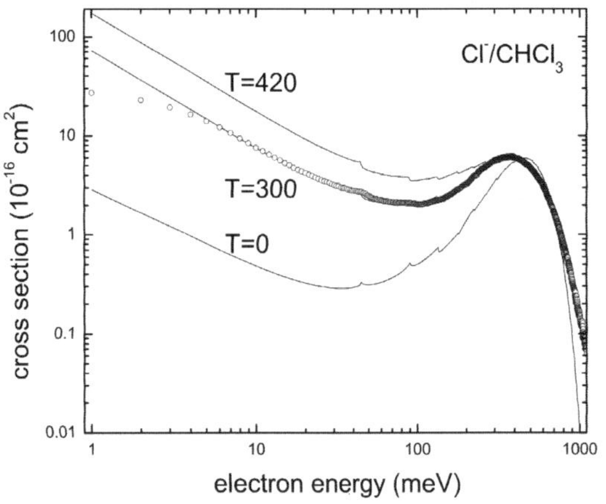

Figure 3. Dissociative attachment to $CHCl_3$ for different vibrational temperatures given in K. Open circles, combined LPA/EXLPA cross sections at room temperature [46]. Both theory and experiment are normalized to swarm measurements.

According to the threshold law for exothermic reactions at very low energies [43], the DEA cross section should be proportional to E^{-x} where the threshold exponent x depends on the molecular dipole moment. For the $CHCl_3$ molecule one obtains $x = 0.623$. As observed in Fig. 3, the vibrationally-averaged R-matrix cross sections are distinctly steeper at very low energies. This can be explained by the presence of a weakly bound or a virtual state due to the combined dipolar and polarization interactions, making the region of validity of the threshold law extremely narrow. For the methyl iodide molecule, for example, the calculated DEA cross section grows much faster towards lower energies than predicted by the threshold law and reaches the correct slope only in the region below 0.1 meV [43]. A similar effect is observed in the present case. However, whereas in the methyl iodide case this effect is caused by a bound state leading to the vibrational Feshbach resonance (VFR) below the first vibrational excitation threshold [15], in the case of chloroform no evidence for a VFR and a bound state is present, therefore the effect is due to a virtual state. We should note also that the discussed behavior is pertinent to collisions with molecules with fixed orientation. In reality the molecules rotate, and when rotation is taken into account, the Bethe-Wigner law $x = 0.5$ should be restored in the sub-meV region [43].

Similar calculations were recently performed for the environmentally important molecules SF_5CF_3 [50] and CF_2Cl_2 [51].

4. Notes on Arrhenius equation

It is well known that the temperature dependence of the rate coefficient for many chemical reactions is described by the Arrhenius equation

$$k(T) = Ae^{-\frac{E_a}{T}}, \tag{2}$$

where T is the temperature in energy units, A is a constant, and E_a is the so-called activation energy. In numerous cases, activation energies have thus been determined from plots of $\ln k$ versus $1/T$. In general, the prefactor A might be temperature-dependent too, but this dependence is normally much weaker than exponential.

Many of the DEA reactions

$$XY + e \to X + Y^-, \tag{3}$$

where X and Y^- are neutral and ionic fragments, not necessarily monoatomic, were found to obey the Arrhenius equation albeit over rather limited temperature ranges. The Arrhenius equation is usually associated with a reaction barrier which might be due to either the endothermicity of the DEA process or due to a barrier separating the initial XY state from the exothermic final $X + Y^-$ state. In the endothermic case we assume that the activation energy is identical with the reaction threshold. In the exothermic case the reaction barrier is usually associated with the energy of the crossing point between the neutral and anion potential curves. The vibrational states of the neutral molecule lying below the energy of the crossing point are expected to have small DEA cross sections. This happens because for these vibrational states at low electron energies the Franck-Condon overlap between the neutral and anion states is small, and for higher energies the electron capture occurs far from the crossing point, and the intermediate anion state is likely to decay before reaching the crossing (stabilization) point. Therefore it is usually assumed that the activation energy is identical with or close to the reaction barrier, i.e. the difference E_B between the energy of the crossing point and the energy of the neutral ground vibrational state.

However, upon a closer inspection of Arrhenius plots for DEA processes [52], one observes that the validity of the Arrhenius equation is certainly limited towards low temperatures and probably towards high temperatures as well. Because of these limitations, the question arises what is the relation between the activation energy and the reaction barrier. We recently performed a detailed study of validity of the Arrhenius law for the DEA process [52]. We were particularly interested in the relation between the activation energy and the reaction threshold (for the endothermic case) or the reaction barrier (for the exothermic case).

The results of this study show that the Arrhenius equation for description of the DEA processes should be used with caution. For endothermic DEA processes and favorable curve crossing, the Arrhenius equation has a rather wide range of validity limited by high temperatures.

The upper temperature limit depends on the threshold energy E_t and becomes quite low for low E_t and low vibrational frequency. The activation energy in the Arrhenius equation is close to E_t.

In the case of exothermic DEA processes with an intermediate barrier the range of validity of the Arrhenius equation is also limited towards low temperatures where the rate coefficient as a function of T becomes flat. The activation energy E_a is found to be lower than the reaction barrier E_B, and the relative difference $(E_B - E_a)/E_B$ decreases towards higher E_B. This behavior is partly due to the relatively slow growth of the partial rate coefficient with rising vibrational energy for vibrational levels just below the barrier. The other reason for the difference between E_a and E_B is the uncertainty in the definition of the reaction barrier: due to the long-range electron-molecule interaction the adiabatic anion curve peaks at lower energies than the diabatic curve. This effect is more significant for stronger long-range contributions to the electron-molecule interaction.

The above conclusions are pertinent to DEA cases where s-wave attachment is allowed. However, low-energy DEA to halogen molecules is dominated by the p-wave. In this case a separate analysis is necessary which was done recently for the Cl_2 molecule [31]. Like in the case of s-wave attachment, the contribution of the ground vibrational state to the rate coefficient is dominant at low temperatures. However, in contrast to s-wave attachment, the p-wave attachment cross section is proportional to $E^{1/2}$ and the rate coefficient proportional to T. Therefore, the temperature dependence in the low-temperature limit is not flat for the p-wave case, nor does it exhibit Arrhenius behavior typical of high temperatures. In fact, the temperature dependence in the low-energy region is substantially weaker than linear because the $E^{1/2}$ dependence of the cross section holds only in the very low-energy region below about 15 meV [22].

These observations are illustrated in Fig. 4 for four molecules: CF_3Br, $CHCl_3$, CH_3Br, and Cl_2. Detailed comparison with experimental data is given in Refs. [24, 31, 32, 46]. Here we just comment on the major features observed in these plots.

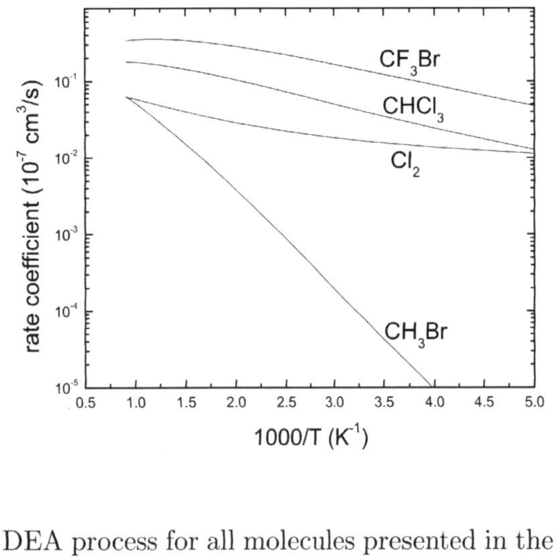

Figure 4. Attachment rate coefficients for CF_3Br, $CHCl_3$, CH_3Br and Cl_2 molecules.

DEA process for all molecules presented in the figure is exothermic. However, due to a rather large reaction barrier, the CH_3Br plot exhibits a substantial activation energy $E_a = 249$ meV which is in a good agreement with swarm measurements [53, 54]. In contrast, the activation energy for CF_3Br and $CHCl_3$ is rather low, and the Arrhenius equation for these species holds in a very limited range of temperatures. This leads to some uncertainties in determination of E_a

and to substantial disagreement between the theoretical and experimental values for activation energy. For example, for CF_3Br [32] the theory predicts $E_a = 52$ meV whereas the experimental value is $E_a = 75$ meV. Another reason for disagreement is a rather slow growth of the theoretical attachment rate coefficient with vibrational temperature, as compared with the experiment. This drawback of the theory might be due to one-mode approximation as discussed in Ref. [32].

5. Attachment to formic acid and glycine

Low energy electron interactions with biological molecules in the gas phase have proved to be a challenge for experimentalists and theorists alike. Although the field has been highly motivated by applications to radiation damage [7, 8, 9], it warrants fundamental interest because of the remarkable variety of bond-breaking reactions produced through the DEA process.

Progress in understanding interactions with large systems such as DNA strands and proteins must be grounded on an accurate analysis of the DEA process in the simplest components of these systems. Surprisingly, in the case of the amino acids and closely related molecules such as formic acid (HCOOH), there is not yet a consensus on the mechanism by which electrons with energies near 1.2 eV produce the sole fragment $(M\text{-}H)^-$, where (M-H) represents the target molecule minus an H atom. We recently presented [55] a theoretical treatment of this process in the two compounds that have been studied most extensively, formic acid and glycine, the simplest of the amino acids. We showed that the existing experimental results can be explained by invoking a single σ^* valence anion state in which the additional electron is largely located on the OH bond [56]. This state has an extremely large width of about 6 eV or higher which precludes its direct observation in the total electron scattering cross section but which is clearly manifested in the DEA [55] and vibrational excitation [57] cross sections at much lower energies. We concluded that in formic acid and glycine, and by implication the other amino acids, capture into the π^* orbital followed by π^*/σ^* coupling, as suggested by Rescigno *et al.* [58], is not required to account for the bond breaking.

To simplify the DEA treatment, we employed again a one-dimensional model according to which the H dissociation occurs along the O-H bond coordinate roughly corresponding to the normal O-H coordinate in the neutral molecule. To find the R-matrix parameters for formic acid, we started with a calculation of the resonance scattering phase shifts using the finite element discrete model (FEDM) method described by Nesbet [59] combined with an approximate version of the variational technique for quasi-bound states developed by Froelich and Brandäs [60].

The R-matrix scattering phase shifts for three values of the O–H distance are presented in Figs. 5 and 6. The resonant contribution is growing so slowly that this growth is completely offset by the dipole contribution. As a result, the total phase shift is decreasing monotonically, although some indication of the resonance contribution appears at the distance $R = 1.2$ Åwhich is closer to the position of the crossing point between the neutral and the anion curves. These features are very similar to those observed for hydrogen halides [17, 18, 19, 61, 62, 63] where the DEA and near-threshold vibration excitation is also dominated by a very broad σ^* resonance.

The corresponding DEA cross section is shown in Fig. 7, together with comparison with experiments [64, 65]. Due to some uncertainties in the fitting procedure, the calculated cross section suffers from some uncertainties as well which were discussed in more detail in Ref. [55]. However all analyzed R-matrix models are able to reproduce the general shape and absolute magnitude of the DEA cross section. We believe that, together with the experimental evidence of Allan [57], this proves the dominant role of the σ^* resonance in excitation of the O–H vibrations and in the near-threshold DEA process.

Similar, albeit semiempirical, calculations [55] were performed for glycine. They reproduce the major features in the DEA cross section measured by Abouaf [66].

Journal of Physics: Conference Series **204** (2010) 012004 doi:10.1088/1742-6596/204/1/012004

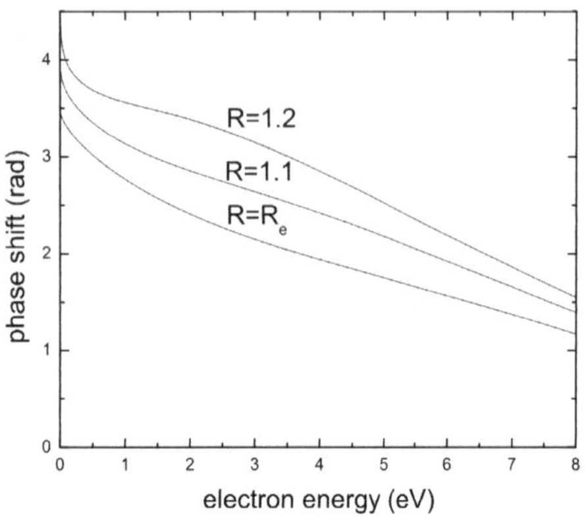

Figure 5. Scattering phase shifts for formic acid for three nuclear geometries.

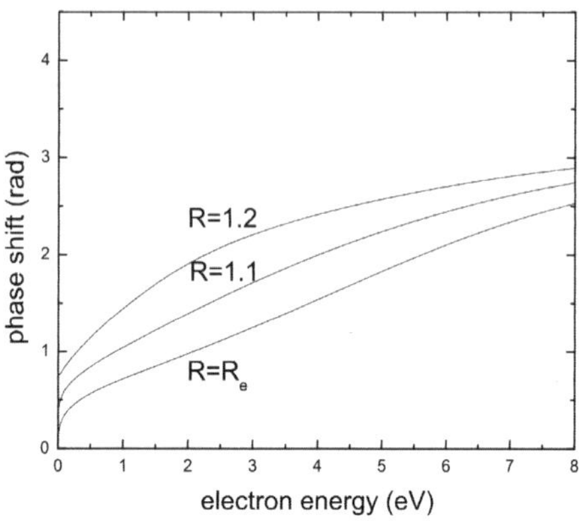

Figure 6. Resonance contribution to the scattering phase shifts for formic acid.

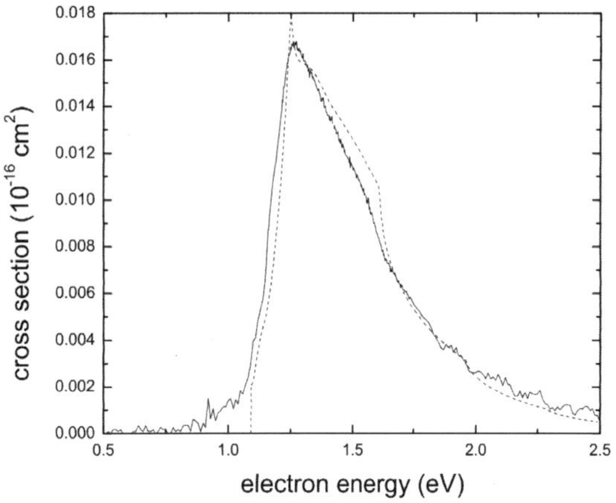

Figure 7. DEA to formic acid. Solid curve, experiment [64, 65]. Dashed curve, theory [55].

6. Conclusion

To summarize, substantial progress has been made during the past five years in our understanding of the basic mechanisms of DEA and in quantitative description of these processes. The resonance R-matrix theory has been successful in explaining major features in the electron energy dependence and vibrational energy dependence of DEA cross sections. Threshold resonances and cusps in DEA cross sections appear to be typical in polyatomic molecules, including molecules of biological interest. In particular we have found similarities in the attachment mechanism for formic acid and hydrogen halides.

Further development of the theory is certainly required. More accurate *ab initio* calculations of the input parameters, resonance positions and widths as functions of nuclear geometry, are necessary. At the same time we need incorporation of multimode effects which still present a big challenge, particularly for the nonlocal version of the resonance theory.

Acknowledgments
The author is thankful to H. Hotop, M.-W. Ruf, P. D. Burrow, G. A. Gallup, and T. M. Miller for fruitful collaboration and to H. Hotop for useful comments on the present manuscript. This work was supported by the US National Science Foundation, Grant No. PHY-0652866, and by Deutsche Forschungsgemeinschaft for visits to TU Kaiserslautern.

References

[1] Christophorou L G (ed.) *Electron-Molecule Interactions and Their Applications* Vols. 1 and 2, (Academic Press, New York, 1984).
[2] Chutjian A, Garscadden A and Wadehra J M 1996 *Phys. Rep.* **264** 393
[3] Christophorou L G and Olthoff J K *Fundamental Electron Interactions With Plasma Processing Gases* (Kluwer Academic/Plenum Publ., New York, 2004).
[4] Toon O B and Turco R P 1991 *Sci. Am.* **264** 68
[5] Lu Q-B and Madey T E 2000 *Surf. Sci.* **451** 238
[6] Lu Q-B and Sanche L 2001 *Phys. Rev. Lett.* **87** 078501
[7] Boudaiffa B, Cloutier P, Hunting D, Huels M A and Sanche L 2000 *Science* **287** 1658
[8] Sanche L 2005 *Eur. Phys. J. D* **35** 367
[9] Surdutovich E and Solov'yov A V 2009 *Europhys. News* **40**, No. 2, 21
[10] Hotop H, Ruf M-W, Allan M and Fabrikant I I 2003 *Adv. At. Mol. Phys.* **49** 85
[11] Domcke W 1991 *Phys. Rep.* **208** 97
[12] Bardsley J N, Herzenberg A and Mandl F 1966 *Proc. Phys. Soc. London* **89** 305; **89** 321
[13] Pearl D M, Burrow P D, Fabrikant I I and Gallup G A 1995 *J. Chem. Phys.* **102** 2737
[14] Wilde R S, Gallup G A and Fabrikant I I 1999 *J. Phys. B* **32**, 663
[15] Schramm A, Fabrikant I I, Weber J M, Leber E, Ruf M-W and Hotop H 1999 *J. Phys. B* **32**, 2153
[16] Wilde R S, Gallup G A and Fabrikant I I 2000 *J. Phys. B* **33**, 5479
[17] Allan M, Čížek M, Horáček J and Domcke W 2000 *J. Phys. B* **33** L209
[18] Čížek M, Horáček J, Sergenton A-Ch, Popović D B, Allan M, Domcke W, Leininger T and Gadea F X 2001 *Phys. Rev. A* **63** 062710
[19] Čížek M, Horáček J, Allan M, Fabrikant I I and Domcke W 2003 *J. Phys. B* **36** 2837
[20] Fedor J, Cingel M, Skalny J D, Scheier P, Märk T D, Čížek M, Kolorenč P and Horáček J 2007 *Phys. Rev. A* **75** 022703
[21] Fabrikant I I, Leininger T and Gadea, F X 2000 *J. Phys. B* **33**, 4575
[22] Ruf M-W, Barsotti S, Braun M, Hotop H and Fabrikant I I 2004 *J. Phys. B* **37** 41
[23] Kolorenč P and Horáček J 2006 *Phys. Rev. A* **74** 062703
[24] Braun M, Ruf M-W, Fabrikant I I and Hotop H 2007 *Phys. Rev. Lett.* **99** 253202
[25] Fabrikant I I 1991 *Phys. Rev. A* **43** 3478
[26] Chourou S T and Orel A E 2008 *Phys. Rev. A* **77** 042709
[27] Tarana M, Wielgus P, Roszak S and Fabrikant I I 2009 *Phys. Rev. A* **79** 052712
[28] Gilbert T L and Wahl A C 1971 *J. Chem. Phys.* **55** 5247
[29] Chutjian A and Alajajian S H 1987 *Phys. Rev. A* **35** 4512
[30] Barsotti S, Ruf M-W and Hotop H 2002 *Phys. Rev. Lett.* **89** 083201
[31] Friedman J F, Miller T M, Schaffer L C, Viggiano A A and Fabrikant I I 2009 *Phys. Rev. A* **79** 032707
[32] Marienfeld S, Sunagawa T, Fabrikant I I, Braun M, Ruf M-W and Hotop H 2006 *J. Chem. Phys.* **124** 154316
[33] Hazi A U, Orel A E and Rescigno T N 1981 *Phys. Rev. Lett.* **46** 918
[34] Bardsley J N and Wadehra J M 1983 *J. Chem. Phys.* **78** 7227
[35] Brems V, Beyer T, Nestmann B M, Meyer H-D and Cederbaum L S 2002 *J. Chem. Phys.* **117** 10635
[36] McCorkle D L, Christophorou L G, Christodoulides A A and Pichiarella L 1986 *J. Chem. Phys.* **85** 1966.
[37] Ingr M, Meyer H-D and Cederbaum L S 1999 *J. Phys. B* **32** L547
[38] Lane A M and Thomas R G 1958 *Rev. Mod. Phys.* **30** 257
[39] Kalin S A and Kazansky A K 1990 *J. Phys. B* **23** 4377
[40] Marienfeld S, Fabrikant I I, Braun M, Ruf M-W and Hotop H 2006 *J. Phys. B* **39** 105
[41] Braun M, Fabrikant I I, Ruf M-W and Hotop H 2007 *J. Phys. B* **40** 659
[42] Gallup G A and Fabrikant I I 2007 *Phys. Rev. A* **75** 032719
[43] Fabrikant I I and Hotop H 2001 *Phys. Rev. A* **63** 022706
[44] Chu S C and Burrow P D 1990 *Chem. Phys. Lett.* **172** 17
[45] Aflatooni K and Burrow P D 2000 *J. Chem. Phys.* **113** 1455
[46] Kopyra J *et al.* 2008 *Int. J. Mass Spectrom.* **277** 130

[47] Kalamarides A, Marawar R W, Durham M A, Lindsay B G, Smith K A and Dunning F B 1990 *J. Chem. Phys.* **93** 4043
[48] Fabrikant I I, Hotop H and Allan M 2005 *Phys. Rev. A* **71** 022712
[49] Braun M, Barsotti S, Marienfeld S, Leber E, Weber J M, Ruf M-W and Hotop H 2005 *Eur. Phys. J. D* **35** 177
[50] Graupner K, Graham L M, Field T A, Mayhew C A, Fabrikant I I, Miller T M, Braun M, Ruf M-W and Hotop H 2008 *Int. J. Mass Spectrom.* **277** 113
[51] Graupner K *et al* 2009 *J. Phys. Chem. A*, submitted
[52] Fabrikant I I and Hotop H 2008 *J. Chem. Phys.* **128** 124308
[53] Wentworth W E, George R and Keith H 1969 *J. Chem. Phys.* **51** 1791
[54] Petrovic Z L and Crompton R W 1987 *J. Phys. B* **20** 5557
[55] Gallup G A, Burrow P D and Fabrikant I I 2009 *Phys. Rev. A* **79** 042701
[56] Scheer A M, Mozejko P, Gallup G A and Burrow P D 2007 *J. Chem. Phys.* **126** 174301
[57] Allan M 2006 *J. Phys. B* **39** 2939
[58] Rescigno T N, Trevisan C S and Orel A E 2006 *Phys. Rev. Lett.* **96** 213201
[59] Nesbet R K 1981 *Phys. Rev. A* **24** 1184
[60] Froelich P and Brandäs E 1975 *Phys. Rev. A* **12** 1
[61] Horáček J and Domcke W 1996 *Phys. Rev. A* **53** 2262
[62] Gallup G A, Xu Y and Fabrikant I I 1998 *Phys. Rev. A* **57** 2596
[63] Čížek M, Horáček J and Domcke W 1999 *Phys. Rev. A* **60** 2873
[64] Pelc A, Sailer W, Scheier P, Mason N J, Illenberger E and Märk T D 2003 *Vacuum* **70** 429
[65] Pelc A, Sailer W, Scheier P and Märk T D 2005 *Vacuum* **78** 631
[66] Abouaf R 2008 *Chem. Phys. Lett.* **451** 25

Journal of Physics: Conference Series **204** (2010) 012005 doi:10.1088/1742-6596/204/1/012005

Electron stimulated desorption of anions containing Oxygen and Nitrogen from self-assembled monolayers of DNA

N Mirsaleh-Kohan[1], A D Bass[1], P Cloutier[1], L Sanche[1,2]

[1]Département de médecine nucléaire et radiobiologie, Faculté de médecine et les sciences de santé, Université de Sherbrooke, Sherbrooke, QC J1H 5N4 Canada
[2]Department of Physics and Astronomy, The Open University, Walton Hall, Milton Keynes, MK7 6AA United Kingdom

Andrew.Bass@Usherbrooke.ca

Abstract. The electron stimulated desorption of anions from self-assembled monolayers of double stranded DNA is reported. Desorption of the oxygen and nitrogen containing anions O^-, OH^-, CN^-, OCN^-, and $OCNH^-$ is induced by the impact of 0.1-20 eV electrons. The anion desorption yields, measured as a function of incident electron energy exhibit pronounced maxima that can be attributed to dissociative electron attachment (DEA) to basic DNA units. Above 15 eV, desorption is attributed to dipolar dissociation (DD). This study further indicates that electrons with energy as low as 2.5 ± 0.3 eV can not only cause damage to DNA but also produce fragments with considerable kinetic energy.

1. Introduction

Although the vast majority of investigations have focused on interactions of high energy radiation with living cells and DNA molecules, it is well established that the genotoxic effects of high energy particles are effectuated by the secondary species generated by the primary particles. These secondary species can cause mutagenic, recombinogenic, and other potentially lethal DNA lesions such as single and double strand breaks [1]. The most abundant of these species are secondary electrons with typical energies below 30 eV [2]. Therefore, understanding the mechanisms of low-energy electron (LEE) induced damage in DNA is highly relevant to radiobiology.

Boudaïfa et al. [3] investigated strand break formation in plasmid DNA by 3-20eV electrons. Single and double strand breaks were observed even at electron energies well below ionization thresholds (7.5 eV to 10 eV) [4]. Thresholds for these breaks were measured to be near 3-5 eV with a maximum at about 10 eV. The authors concluded that these breaks occur as a result of electron attachment to DNA subunits to form a transient negative ion (TNI), and subsequent bond dissociation. The threshold for strand-breaks in their work differs from that for damage induced by photons, where the onset lies at about 7 eV [5].

To better understand the mechanisms underlying DNA damage through electron attachment, studies have examined interactions of LEE with various components of DNA such as the bases [6-10], deoxyribose analogues [11,12], and the phosphate group [13] of the backbone in gas and solid phases. For all of the DNA components [6-13], the formation of resonances (i.e., TNI) that dissociate into a stable anionic and radical fragments was observed. One of the remaining challenges is to identify, in a long DNA chain, which TNI state of the different basic constituents (the bases, sugar and phosphate group and structural water) lead to specific damages (e.g. which TNI on what basic units lead to DNA strand breaks).

Pan et al. [14] investigated the electron stimulated desorption (ESD) of anions induced by 3-20 eV electron impact on lyophilized thin films of linear and supercoiled DNA. Their results suggested that dissociative electron attachment (DEA) is the process that leads to the formation of the observed desorbed anions (H^-, O^-, and OH^-).

Techniques developed to form self-assembled monolayers (SAMs) of alkanethiols on gold substrates [15] are also applicable to thiolated DNA [16]. The high order and purity of the resulting DNA SAMs offer distinct advantages over other techniques (such as lyophilization) when preparing thin film samples of DNA. In this paper we report the electron stimulated desorption (ESD) of anions from SAM films of double stranded thiolated DNA on clean gold substrates using 40-mer oligonucleotides. A long chain of DNA was used to increase the complexity of our DNA model. Anion desorption was analyzed as a function of incident electron energy (0-20 eV) and irradiation time at a fixed electron energy employing a high-sensitivity time-of-flight mass spectrometer. This study provides information on the TNI states responsible for damage to specific basic units of DNA.

2. Experimental

Phosphothiolated oligonucleotides of double stranded DNA were prepared with the 40-base sequence 5'-GGTAC, CAGGC, CTACT, ACGAT, TTACG, AGTAT, AGCGA, GCTCG-3', where G stands for guanine, T for thymine, A for adenine, and C for cytosine (see Figure 1). The thiolated oligonucleotide and its complimentary strand were purchased from University Core DNA services at the University of Calgary. Thiolation was performed by substitution of five oxygens doubly bonded to phosphorus with sulfur atoms (5S-dsDNA). The 5S-dsDNA is assumed to lie parallel to the surface [17] as shown in Figure 1. Gold plates were used as substrates and purchased from Arrandee, Germany. They were cleaned, prior to chemisorption of the DNA, by multiple exposure to UV/ozone and rinsing with ethanol and doubly-distilled, deionized water (pure water).

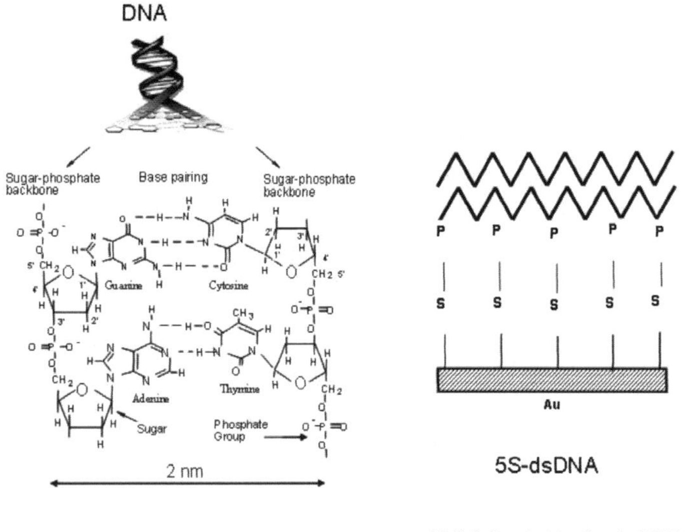

A) Part of DNA structure

B) 5-Sulfur double-stranded DNA chemisorbed on a gold substrate

Figure 1 (a) Part of double stranded DNA (b) Schematic drawing of self-assembled monolayer (SAM) of double stranded DNA chemisorbed on a gold substrate

SAM DNA films were prepared by immersing the cleaned gold substrate in the DNA solution [18] for 12 hours. The plates then were immersed in pure water three times for 20 minutes to remove any unbound molecules. Finally, the plates were dried under a stream of N_2 gas and placed into a load-lock vacuum system ($<1\times10^{-8}$ torr) to facilitate transfer into the ultrahigh vacuum chamber ($\sim5\times10^{-10}$ torr) for LEE bombardment.

Anions desorbed by LEE impact were measured by a time-of-flight (TOF) mass analyzer. The ESD system and its specifications have been discussed in detail previously [19]. Recently, a newly designed load-lock has been coupled to the TOF mass analyzer, and permits us to introduce and to study samples prepared under ambient conditions. After evacuation for a minimum of about 4 hours, the films can be transferred, sequentially into the main chamber for irradiation.

The DNA SAM films were bombarded by LEE from an electron gun (Kimball Physics ELG-2). The beam was at an incident angle 45° with respect to the substrate and focused into an estimated spot-size of $1 mm^2$. The electron gun was operated in a pulsed mode with 800 ns duration at a rate of 5 kHz. The films were irradiated from 0.1 to 20 eV with 4nA time-averaged transmitted current measured at 10 eV. The energy scale of the incident electron is estimated to have uncertainty of ±0.3 eV and the full-width, half-maximum (FWHM) of the electron energy is approximately 0.5 eV. A short time (~10 ns) after the electron beam pulse, desorbed anions were repelled from the films by a negative potential pulse (rise/fall time 30 ns, pulse width of 2 μs) applied to the gold substrate, directing the anions into the entrance optics of a reflectron TOF mass analyzer positioned at a distance of 10mm from the film surface.

Three series of ESD experiments were performed: (a) the mass spectrum of each SAM film was measured at a fixed electron energy (b) the variation of the desorbed ions signal was monitored as a function of incident electron energy to obtain the yield function for each desorbed ion, and (c) the time-dependence of ion signal was recorded at a fixed incident electron energy.

To examine the efficiency of the DNA adsorption on the gold substrates, X-ray Photoelectron Spectroscopy (XPS) was performed using our previously described XPS spectrometer [20]. The X-ray source (15kV, 350 W) uses an Al anode. The X-ray beam impinged onto the sample surface at an incident angle of 72° relative to the substrate while the hemispherical electron energy analyzer was positioned normal to the sample surface. The pass energy was set to 11.75 eV which results in a resolution of between 2 to 3 eV. XPS spectra for Nitrogen, N1s (binding energy of ~400 eV) and Phosphorous, P2p (binding energy of ~133 eV) peaks were recorded for a 5S-dsDNA film as shown in Figure 2.

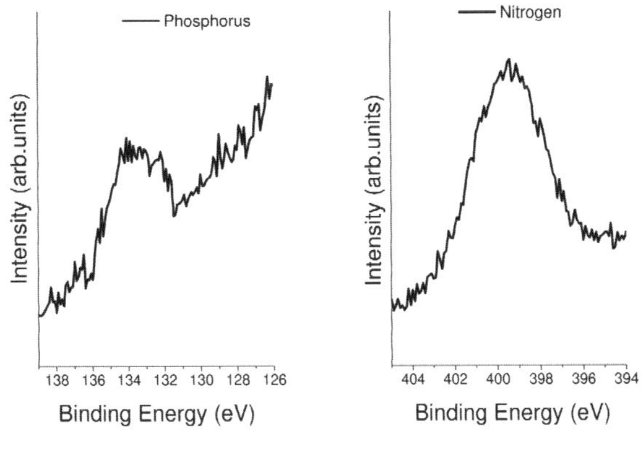

Figure 2 XPS measurements of N1s (binding energy of ~400e V) and P2p (binding energy of ~133eV) peaks for a 5S-dsDNA film where intensity is proportional to the amount of DNA chemisorbed on a gold substrate.

3. Results

We report in this section the desorption of the anions O^-/NH_2^- (16amu), OH^- (17amu), CN^- (26amu), OCN^- (42amu), and $OCNH^-$ (43amu) induced by LEE impact onto 5S-dsDNA SAM films. No anions more massive than 43amu were observed. This may in part be due to heavier fragments tending to possess insufficient kinetic energy to overcome attractive polarization and charge-image forces induced in the film and metal substrates by their negative charge [21].

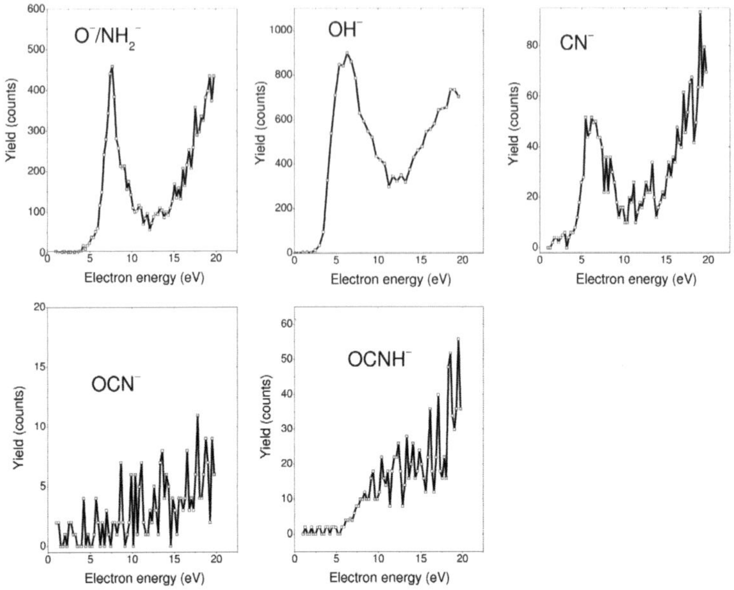

Figure 3 Incident electron energy dependence of desorbed anions yields from a 5S-dsDNA SAM

The dependence of the desorbed anion yields on incident electron energy (*yield functions*) for a 5S-dsDNA SAM is shown in Figure 3. The anion yield functions exhibit energy thresholds of 4-5 eV, except for OH^- which displays a threshold closer to 2.5 eV. Most yield functions show resonance structure near 7-8.5 eV, but OH^- has a maximum resonance near 6.5 eV. The yield function for OCN^- does not show as strong a signal, threshold or resonance maximum as can be seen for the other anions. However, irradiation for an extended period of time [22] shows that this ion has a threshold of about 5 eV and a maximum around 6.7 ±1 eV. All anion signals show a monotonic increase at energies above 14-15 eV. Variations of the desorbed anion signals as a function of time, at incident energy of 8 eV were also recorded and are reproduced in Figure 4 for anions O^-, OH^-, CN^-, OCN^- and $OCNH^-$.

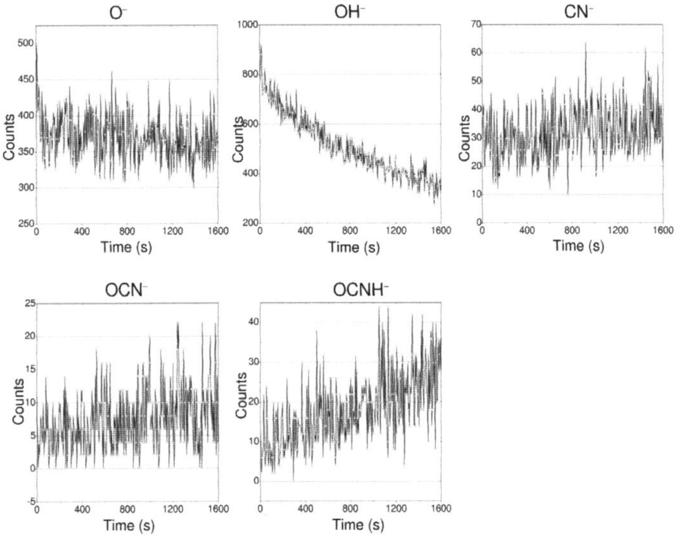

Figure 4 Time dependence of desorbed anions from a 5S-dsDNA film recorded at incident electron energy of 8 eV.

4. Discussion

Formation of a stable negative ion by LEE impact usually occurs through DEA for electron energies less than 15 eV and/or dipolar dissociation (DD) for electron energies above 15 eV. In the DEA process, a TNI dissociates into a stable negative ion and one or more neutral species. Alternatively, in the case of large molecules, dissociation of a TNI can break a bond leaving the excess electron on one side of the broken bond and forming a radical on the other side. Since the initial step in the DEA process is electron attachment to form a TNI, the yield function of the anion fragment or bond breaking exhibits pronounced maxima at the resonance energy. A negative ion can also be produced through the DD process where the ion pair is a result of dissociation of an electronically excited molecule. In the DD process, the anion yield function always has a non-resonant behavior to which can be superimposed a resonant line shape; the non-resonant portion of the signal increases monotonically with electron energy. Anion desorption may also arise from the reactive scattering [23] of a primary anion (produced by for example, DEA) with surrounding molecules. Thus, the yield function of the secondary anion usually carries the resonance signature of the primary ion yield, but with a much smaller magnitude. Stable anions may also be formed through other reactions of primary anions, but such species do not posses sufficient energy to desorb. However, any new chemical species synthesized during electron impact may also undergo DEA and DD and contribute to the desorption signal, but this may only become apparent at long irradiation times.

In the present experiments, the energy dependence of all desorbed anions shows a resonance feature around 7-8.5 eV that is a typical of DEA. The continuous rise in the anions signals above 15 eV is typical of the non-resonant, dipolar dissociation process. Thus interaction of LEE with DNA SAMs in our experiment produces fragment anions via both DEA to DNA and DD as previously reported [14]. The time dependence of anion desorption signals at a fixed incident electron energy reflects the processes responsible for the production of each anion. In the case of DEA and reactive scattering from intact molecules, the anion desorption signal is expected to decrease exponentially with time, whereas the signal deriving from molecules synthesized by the electron beam should increase as a function of time. As shown in Figure 4, time dependence of O^-/NH_2^- shows an exponential decay. The mass resolution of our apparatus does not allow us to differentiate between O^- and NH_2^- for mass 16amu. However, more detailed investigation on the latter revealed that 16amu desorbed from SAM films is more likely to be oxygen [22, 7]. In the ESD experiment presented here, O^- appears to arise from DEA of DNA components since the time dependence of this ion shows an exponential decay. Previously, Pan et al.[14] concluded that desorption of O^- signal from linear and supercoiled DNA results of DEA to the phosphate group of DNA, more specifically from P=O bond of the phosphate group. Furthermore, electron-stimulated desorption of O^- from thin films of phosphate film (NaH_2PO_4) was investigated [13]. In this latter experiment, the yield function of oxygen anion presented a single broad peak at 8.0 eV and a threshold of 4.7 eV in good agreement with the threshold (5 eV) and maximum (7.9 eV) seen in our experiment (the difference in energies is within our experimental uncertainty). Thus, the signal of O^- in our ESD experiment is most likely produced from DEA to the phosphate group of DNA.

The OH^- yield function exhibits a threshold of about 2.5 eV with a maximum around 6.5 eV (Figure 3). The time dependence of OH^- signal exhibits an exponential decay with time as shown in Figure 4, indicating that the observed OH^- desorption arises from DEA to the DNA and/or a reactive scattering processes. In the case of reactive scattering, OH^- can be generated via reactive scattering of O^-, with for example, hydrocarbon species [24]. However the present OH^- yield function does not greatly resemble that of O^- and is of much higher intensity, indicating that OH^- ions do not originate from reactive scattering of O^-. We thus attribute the OH^- ions signal to DEA and DD to the DNA components.

Anions containing nitrogen, CN^-, OCN^-, and $OCNH^-$ in the present experiment show very weak signals compared to other observed anions. This weakness may indicate that these anions are unable to easily escape from the surface due to higher mass and their (presumed) low kinetic energies. Nitrogen-based fragment anions necessarily originate from the DNA bases, since only bases contain nitrogen.

Abdoul-Carime at al. [7] investigated nitrogen-containing anions desorbed from solid films of the four DNA bases. In each case, the cyanided anion, CN^-, was observed to desorb, while OCN^- desorption was only observed from thymine films. Interestingly, no $OCNH^-$ was detected from any of the bases. Several studies have also investigated fragmentation of gas-phase DNA bases by LEE [6, 8, 9]. In the gas phase, CN^- was detected from all DNA bases [6, 8, 9], whilst OCN^- was observed from only guanine, thymine and cytosine [6, 8, 9]. $OCNH^-$ was only detected from fragmentation of thymine [8]. CN^- and OCN^- were also observed to desorb from thin films of short oligonucleotide GCAT by impact of LEE [25]. The time dependence of these anions in Figure 4 shows a slightly increase as a function of irradiation time, suggesting that these ions are generated from LEE-induced species. This hypothesis could also partially explain why the signal from these anions is initially so low.

In summary, LEE (0.1-20 eV) impact on 5S-dsDNA films produces the anion fragments O^-, OH^-, CN^-, OCN^-, and $OCNH^-$ as shown. From the dependence of the desorbed anion signals on incident electron energy, it is seen that the desorbed species arise from direct dissociative electron attachment to- and dipolar dissociation of DNA basic units. In our experiment, mass 16amu can be attributed to the oxygen ion and arises as a result of DEA to phosphate group of DNA. Nitrogen-containing ions, which have been observed previously in ESD experiments from thin films of the DNA bases, were here observed to also desorb from a long DNA chain (40-mer). Thus it is likely that damage processes observed for the pure bases remain effective once the molecules are bonded with the sugar in DNA. However the increase in signal with irradiation-time suggests that the yields of these species are greater from electron degraded bases.

Finally, according to our ESD results, desorption of anions induced by LEE occurs at a threshold energy as low as 2.5 eV (OH^-) in a good agreement with the results on single and double stranded breaks of DNA [3]. Here, electrons with energy as low as 2.5 eV can induce anion fragments from DNA with sufficient kinetic energy (~ 1eV) to escape the induced polarization potential. This threshold is well below that reported for the DNA damage induced by UV irradiation [26] emphasizing the different damage mechanisms induced by LEE and photons. Furthermore, one can speculate that following DEA, the radical left on the DNA could lead to a strand break while reactive scattering by the dissociating anions with the other strand might produce further damage. Although the probability of such subsequent processes is expected to be low, it nevertheless offers a mechanism for double strand breaks with electrons of energy as low as 2.5 eV.

References

[1] For a review see Sanche L 2005 *Eur. Phys. J. D.* **35** 367 and references therein
[2] Laverne J A and Pimblott S M 1995 *Radiat. Res.* **141** 208
Cobut V, Frongillo Y, Patau, J P, Goulet T, Fraser M-J and Jay-Gerin J-P 1998 *Radiat. Phys. Chem.* **51** 229
[3] Boudaïfa B, Cloutier P, Hunting D, Huels M A and Sanche L 2000 *Science*, **287** 1658
[4] Colson A O, Besler B and Sevilla M D 1992 *J. Phys. Chem.* **96** 9787
[5] Hieda K 1994 *Int. J. Radiat. Biol.* **66** 561
[6] Abdoul-Carime H, Langer J, Huels M A and Illenberger E 2005 *Eur. Phys. J. D.* **35** 399
[7] Abdoul-Carime H, Cloutier P and Sanche L 2001 *Radiat. Res.* **155** 625
[8] Huels M A, Hahndorf I, Illenberger E and Sanche L 1998 *J. Chem. Phys.* **108**, 1309
[9] Denifl S, Ptasińska S, Probst M, Hrušák J, Scheier P and Märk T D 2004 *J. Phys. Chem. A* **108** 6562
[10] Hervé du Penhoat M-A, Huels M A, Cloutier P, Jay-Gerin J-P and Sanche L 2001 *J. Chem. Phys.* **114** 5755
[11] Antic D, Parenteau L, Lepage M and Sanche L 1999 *J. Phys. Chem. B.* **103** 6611
[12] Antic D, Parenteau L and Sanche L 2000 *J. Phys. Chem. B.* **104** 4711
[13] Pan X and Sanche L 2006 *Chem. Phys. Lett.* **421** 404
[14] Pan X, Cloutier P, Hunting D and Sanche L 2003 *Phys. Rev. Lett.* **90** 208102
[15] Swalen J D, Allara D L, Andrade J D, Chandross E A, Garoff S, Israelachvili J, McCarthy T J,

Murray R, Pease R F, Rabolt J F, Wynne K J and Yu H 1987 *Langmuir* **3** 932
[16] Herne T and Tarlov M 1997 *J. Am. Chem. Soc.* **119** 8916
[17] Abdoul-Carime H, Dugal P-C and Sanche L 2000 *Radiat. Res.* **153** 23
[18] Pan X and SancheL 2005 *Phys. Rev. Lett.* **94** 198104
[19] Hedhili M N, Cloutier P, Bass A D, Madey T E and Sanche L 2006 *J. Chem. Phys.* 125 94704
[20] Klyachko D V, Huels M A and Sanche L 1999 *Radiat. Res.* **151** 177
[21] Sambe H, Ramaker D E, Parenteau L and Sanche L 1987 *Phys. Rev. Lett.* **59** 236
[22] Mirsaleh-Kohan N, Bass A D, Cloutier P and Sanche L *Unpublished*
[23] Bass A D and Sanche L *2003 Low Temp. Phys.* **29** 202
[24] Bass A D, Parenteau L, Huels M A and Sanche L 1998 *J. Chem. Phys.* **109** 8635
[25] Ptasinska S and Sanche L 2006 *J. Chem. Phys.* **125** 144713
[26] Folkard M, Prise K M, Vojnovic B, Brocklehurst B and Michael B D 2000 *Int. J. Radiat. Biol.* **76** 763

Journal of Physics: Conference Series **204** (2010) 012006 doi:10.1088/1742-6596/204/1/012006

Three-Body Effects in Dissociative Recombination of Molecular Ions

Rainer Johnsen

Department of Physics and Astronomy, University of Pittsburgh, Pittsburgh, PA 15260

rj@pitt.edu

Abstract. This contribution discusses possible effects of the plasma environment on dissociative recombination (DR) of two ion species, $H^+(H_2O)_4$ and H_3^+, for which afterglow measurements give larger recombination coefficients than beam experiments. This raises the question to what extent the apparent discrepancies reflect differing recombination mechanisms or are simply due other "errors" or "plasma complications". $H^+(H_2O)_4$ recombination seems to be a case in which third-body effects are very strong. It also has been suggested that afterglow H_3^+ recombination coefficients can be reconciled with the lower storage-ring and theoretical results by invoking third-body stabilization of initially formed high Rydberg states, but the estimates presented here indicate that one should be cautious in accepting such conclusions.

1. Introduction

Experimental and theoretical studies of dissociative recombination (DR) of molecular ions with slow electrons,

$$ABC^+ + e^- \rightarrow neutral\ fragments, \tag{1.1}$$

have been pursued for about 60 years but there are still some nagging discrepancies. I begin by pointing out that Eq. (1.1) can be read in two ways, either as representing a binary collision or as a reaction equation that leaves unspecified the degree to which the "environment", e.g. neighboring charged or neutral particles, influence the rate and products of the reaction. In the afterglow technique, the decay of the electron density in a plasma is measured subsequent to a brief ionizing pulse. Here, the neutral gas density (roughly from 10^{16} to 10^{19} cm^{-3}) and electron density (10^9 to 10^{10} cm^{-3}) tend to be fairly high. In recent years, experimental work on DR has shifted to merged-beam experiments, in particular ion storage rings, that yield accurate binary recombination cross sections and branching fractions with very high energy resolution. In many cases, for instance those in which diatomic ions recombine by the fast direct mechanism, the thermal rate coefficients derived from storage-ring data agree rather well with the older afterglow data. However, there are exceptions. The question is not which of the techniques gives the "correct" rate coefficient but rather which coefficient might be appropriate for a given application.

I will examine two examples: The first deals with a big "floppy" ion, a hydrated proton, $H^+ (H_2O)_4$ (also written as $H_3O^+ (H_2O)_3$). I chose this ion species because there are good data from both afterglows and ion storage rings and because of its importance in the D-region of the terrestrial ionosphere. Recent storage-ring experiments [1] indicate that fragmentation into $4\ H_2O + H$ is the dominant dissociation channel, but they leave open the question to what extent dissociation occurs by a sequential process. The total energy released in this channel is moderate, only about 3.4 eV, and much of that energy release is likely to be converted to internal rather than kinetic energy of the fragments. In the case of an ion with numerous internal states it seems plausible that capture of an electron results in a fairly long-lived complex that can be collisionally stabilized, similar to the well known mechanism of electron attachment to SF_6.

The second example concerns the recombination of H_3^+, the most abundant (and most studied!) molecular ion in the universe. It plays a crucial role in the formation of molecular species in interstellar clouds. The long (and often controversial) history of H_3^+ has seen several attempts to reconcile discrepant afterglow and beam results and to "close the books" on the subject. Invoking third-body effects constitutes just one of these attempts. While it would be most gratifying to dispose of the remaining discrepancies, my estimates to be discussed below indicate that the proposed third-body mechanism fails to explain the magnitude of the observed increase of the recombination rate with neutral density.

2. Recombination of $H^+(H_2O)_4$

Afterglow measurements for the n=4 cluster are available over a wider range of gas densities than for other cluster sizes of the same series. The equilibrium constants governing the relative concentrations of different cluster sizes are such that it is fairly easy to produce afterglow plasmas in which the n=4 cluster dominates and to obtain good data. Fortunately, we now also have data [1] from the storage rings on total rate coefficient and product branching fractions.

Figure 1 shows a comparison of the afterglow data [2 ,3 ,4] (all at 300K), as function of the neutral gas density (helium or neon) to the 300 K value inferred from the storage-ring cross sections. The observed rate coefficients initially increase with density but then approach a nearly constant value.

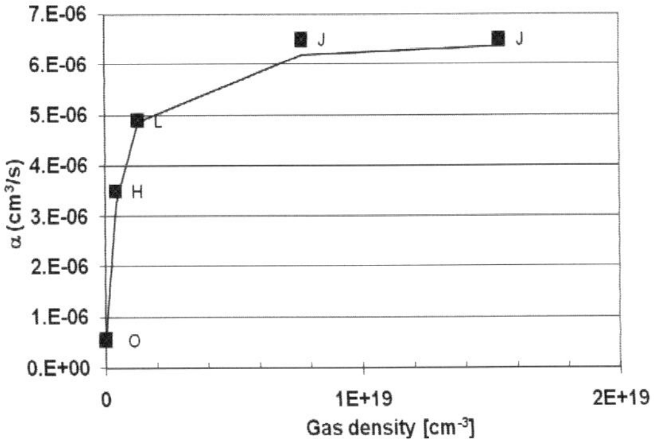

Figure 1.
Measured (T~300 K) recombination coefficients for $H^+(H_2O)_4$, as a function of experimental neutral gas density. Data labels and references: O: storage ring [1], H: afterglow [2], L afterglow[3], J: high-pressure afterglow [4].
Line: Fit using Eq. (1.2).

For other cluster sizes the storage ring data also give lower rate coefficients than the afterglows, but the density effect is less pronounced, and there are no good afterglow data at very high densities.

We do not yet have a clear theoretical understanding why cluster ions recombine very rapidly. In the picture outlined by Bottcher [5] the captured electron becomes temporarily bound by exciting one or more vibrational modes and intramolecular energy transfer then renders the collision complex relatively long-lived against auto-ionization. This lead to the possibility that in the presence of third-body atoms, part of the energy of the complex may be transferred to the third body, thus inhibiting auto-ionization while leaving more time for dissociation.

A simple quasi-equilibrium model suggests that the overall recombination coefficient should be written as the sum of the binary and the three-body channel, i.e.

$$\alpha = \alpha_1 + \alpha_2 \frac{k_s \tau [M]}{1 + k_s \tau [M]} \tag{1.2}$$

where τ is the complex lifetime, M the concentration of the third bodies, and k_s is a rate coefficient that describes collisional stabilization of the complex. In the limit of high M, the rate coefficient is limited by capture rate of electrons. Equation (1.2) reproduces the experimental observations quite well for $\alpha_1 = 5.5 \times 10^{-7}$ cm^3/s, $\alpha_2 = 6 \times 10^{-6}$ cm^3/s, and taking the product $\tau\, k_s$ as ~2×10^{-18} cm^3. If one estimates $k_s \sim 10^{-9}$ cm^3/s, the complex lifetime would have to be on the order of 2 ns, which does not seem unreasonably long. In spite of the apparent success of this crude model, it is clearly no substitute for rigorous theory. I only wish to make the point that one should not too quickly relegate the older afterglow data to the "dust bin of history". The storage ring data may be exactly what is needed to model very thin media like interstellar clouds, but the afterglow data may be more appropriate for models of denser plasmas.

3. Three-body effects in the DR of H_3^+?

I will skip a discussion of the rather convoluted history of H_3^+ studies since I reviewed it not too long ago [6]. At this time theoretical and experimental results strongly favor acceptance of a moderately large 300 K binary recombination rate coefficient of about 7-8 × 10^{-8} cm^3/s.

However, most afterglow experiments (if sufficient hydrogen is present) give rates that are higher by about a factor of two or more. Although the data are quite scattered, there is a roughly linear increase with the neutral densities, compatible with a three-body coefficient of about 3×10^{-25}/cm^6/s. Glosik et al [7] show a graph of the data and propose a three-body mechanism that expands on an earlier conjecture by Gougousi et al. [8]. The recombination is thought to involve resonant capture of low-angular momentum electrons into high Rydberg states, followed by *l*-mixing in collisions with neutral atoms, which inhibits auto-ionization and increases stabilization.

Does such a mechanism really explain the observed density dependence and thus reconcile afterglow and beam data?

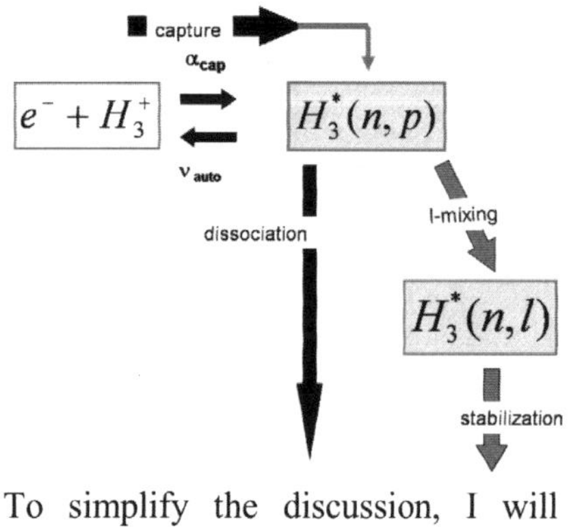

Figure 2 Electron capture and *l*-mixing scheme. Boxes labeled $H_3^*(n,p)$ represent high Rydberg states with principal quantum numbers n in *p*-states. $H_3^*(n,l)$ denotes statistical mixtures of all possible *l*-states.

To simplify the discussion, I will assume that only *p*-states are formed initially. Binary recombination occurs by predissociation of the H_3^* *(n,p)*, while the three-body mechanism is thought to proceeds *via* *l*-mixing, and subsequent "stabilization", which includes any process that irreversibly destroys $H_3^*(n,l)$. *l*-mixing is an essential part of the scheme since it multiplies the statistical weight by a large factor (about of 1000 for a Rydberg population from n=25 to 80) and hence the stabilization step can be relatively slow. The efficiency of the three-body mechanism may be estimated if one applies detailed balancing to obtain the equilibrium constant

$$K_{eq}(p) = \frac{H_3^*(n,p)}{[H_3^+] n_e} = \alpha_{cap} \tau_a \tag{1.3}$$

Here, α_{cap} represents the rate of electron capture into *p*-states of high Rydbergs. The three-body recombination coefficient is then limited by the value

$$\alpha_3 = K_{eq}(p)k_l \tag{1.4}$$

where k_l denotes the rate coefficient for inducing *l*-mixing in collisions with the neutral atoms (helium in most cases). Glosik et al [7]estimate $K_{eq}(p) \sim 3.5 \times 10^{-17}$ cm^3 from theoretical average lifetimes $\tau_a \sim 100$ ps for a range of Rydberg states with *n* from 25 to 80, and an estimated value of $\alpha_{cap} \sim 3.5\times10^{-7}$ cm^3/s (Note: a slightly lower estimate for $K_{eq}(p)$ can be obtained from the Saha equation).

To reproduce the (very approximate) experimental value $\alpha_3 \sim 3\times10^{-25}$/cm^6/s using Eq. (1.4) one needs mixing rate coefficients k_l on the order of 10^{-8} cm^3/s or larger. The problem is that such large *l*-mixing coefficients in rare gases are compatible with theory [9] and experiment only for Rydberg states near *n*=10. For higher *n* the rate coefficients fall off rapidly as $n^{-2.7}$. Thus I conclude that *l*-mixing for the high n-states (25 to 80) cannot be fast enough to explain the observed three-body rates. One should keep in mind, however, that an n=80 Rydberg state is very large! At a pressure of 1 Torr, several thousand atoms reside within its volume making the use of *l*-mixing rate coefficients somewhat dubious! It is also possible that at high ambient electron density *l*-mixing due to plasma electrons (those mixing cross sections are extremely large and increase as n^5) may speed up the mixing, but there is a further problem. Even if complete *l*-mixing were to occur, one still needs a mechanism to "stabilize" the *l*-mixed population. A gradual removal of energy by *n*-reducing collisions with neutral atoms encounters the same "bottleneck" that makes three-body recombination of atomic ions very slow. Stabilization by predissociation may help, but its efficiency is hard to estimate because we do not know the needed predissociation rates for the many states involved.

4. Conclusions

While experimental findings suggest that some dissociative recombination processes are enhanced by the plasma environment, a satisfactory theoretical picture has not yet emerged. The question is not purely "academic" and should be pursued further. Realistic modelling of plasmas may involve more than the purely binary recombination mechanism!

References

[1] Öjekull, J, Andersson, P U, Pettersson, J B C, Marković, N, Thomas, R D, Al Khalili, A, Ehlerding, A, Österdahl, F, Af Ugglas, M, Larsson, M., Danared, H and Källberg, A 2008, *J. Chem. Phys.* **128**, 044311

[2] Huang, C-M, Whitaker, M, Biondi, MA, and Johnsen, R 1978, Phys. Rev. A **18**, 64

[3] Leu, M T, Biondi, M A and Johnsen, R 1973, *Phys. Rev. A* 7, 292

[4] Johnsen, R 1993, *J. Chem. Phys.* **98**, 5390

[5] Bottcher C,1978, *J. Phys. B* **22**, 3887

[6] Johnsen, R 2005, *J. Phys. Conf. Ser* **4**. 83-91

[7] Glosik, J, Korolov, I, Plasil, R, Novotny, O, Kotrik, T, Hlavenka, P, Varju, J, Mikhailov, I A, Kokoouline, V and Greene, C H2008, *J. Phys B* **41,** 191001

[8] Gougousi, T Johnsen, R and Golde, M F 1995, *Int. J. Mass Spectr. Ion Proc,* **149**, 131

[9] Hickman, A P 1978, *Phys. Rev.* A **18**, 1339

Journal of Physics: Conference Series **204** (2010) 012007 doi:10.1088/1742-6596/204/1/012007

Calculation to a prescribed accuracy of the transport cross sections for collisions between structureless particles

Larry A. Viehland and Yongbin Chang
Science Department, Chatham University, Pittsburgh, PA 15232
E-mail: ychang@chatham.edu

Abstract. Careful analysis of multiple orbiting for collisions in classical mechancis shows that a larger impact parameter does not necessarily result in a larger orbiting separation. To take such possibilities into account, a new computer program has been developed for calculating the transport cross sections when a potential energy curve is supplied as a table of points. It is shown that the new program is superior to those used previously, but that it is not necessary to redo results obtained over the past 25 years in situations where transport cross sections were obtained to the requested accuracy by the previous programs.

1. Introduction

There are several methods [1] for determining gaseous ion transport coefficients over wide ranges of the gas temperature, T_0, and the ratio, E/n_0, of the electric field strength to the gas number density. Similar methods are involved in determining the transport properties of neutral gases as functions of T_0 [2]. All of these methods depend upon calculating a set of transport cross sections, $Q^{(l)}(E)$, over wide ranges of the kinetic energy, E, for collisions between two structureless particles and for many $l \geq 1$. The first two cross sections are known as the momentum-transfer (or diffusion) and viscosity (or thermal conductivity) cross sections, respectively, but the others are unnamed because they arise only in corrections to the first-approximation formulas that are dominated by $Q^{(1)}(E)$ and $Q^{(2)}(E)$.

The transport cross sections are defined by the classical-mechanical expression [1]

$$Q^{(l)}(E) = 2\pi \int_0^\infty \left[1 - \cos^l \chi(b, E)\right] b db, \tag{1}$$

or by an equivalent expression using Legendre polynomials of the cosine rather than powers. Here b is the impact parameter and χ is the deflection angle defined by the expression [1]

$$\chi(b, E) = \pi - 2b \int_{r_m}^\infty \frac{dr}{r^2 \; [F(r, b, E)]^{1/2}}. \tag{2}$$

The lower limit on the integral in eq. (2) is the classical turning point, i.e., the largest, positive, real root of

$$F(r, b, E) = 1 - \frac{b^2}{r^2} - \frac{V(r)}{E} = 1 - \frac{V_{eff}(r)}{E}. \tag{3}$$

It is through eq. (3) that the position-dependent interaction potential, $V(r)$, affects the transport cross sections and hence the transport coefficients.

The $Q^{(l)}(E)$ are single-valued functions of E for each integer value of l. Following Smith and coworkers [3, 4], we adopt the condition that we must be able to specify a fractional error, ϵ, such that the final values of the $Q^{(l)}(E)$ are known to be correct within this relative accuracy. Calculation difficulties arise because of singularities at or near the endpoints of the intervals of integration in eqs. (1) and (2), particularly when E, b and r are such that the colliding particles can orbit about one another indefinitely. These singularities are best understood in terms of the effective potential

$$V_{eff}(r) = V(r) + \frac{Eb^2}{r^2}. \tag{4}$$

Orbiting can only occur at energies smaller than some critical value, E_c, that depends upon the interaction potential. This is because larger values of E mean that the $V(r)$ term in eq.(4) is negligible compared to the other term. Eq. (3) shows that, for orbiting to occur, there are three conditions that must be satisfied by the effective potential at an orbiting energy $E_o < E_c$, an orbiting separation, r_o, and an orbiting impact parameter, b_o. They are $V_{eff}(r_o) = E_o > 0$, $V'_{eff}(r_o) = 0$, and $V''_{eff}(r_o) < 0$. The second condition can be shown from eq. (4) to mean that $E_o b_o^2 = \frac{1}{2} r_0^3 V'(r_o)$. Hence the first and the third conditions become $E_o = V(r_o) + \frac{1}{2} r_o V'(r_o) > 0$ and $V''(r_o) + 3r_o^{-1} V'(r_o) < 0$.

Orbiting difficulties are further complicated by the phenomenon of multiple orbiting discovered by Rainwater *et al.* [5]. Here there are two or more pairs, (b_o, r_o), for a particular orbiting energy. Multiple orbiting was considered by one of the present authors shortly after its discovery [6, 7], in the context of numerical rather than functional potentials, i.e., in the situation where the interaction potential is supplied as a tabulated set of points, (V, r). However, it has been assumed in all of the work on multiple orbiting to date that the orbiting impact parameter always increases when the orbiting separation increases. While this is ordinarily the case, we have found situations for numerical potentials where the reverse is true over small regions of orbiting energy. The exceptions can be understood by comparing the derivatives of b_o and E_o with respect to r_o; it is found that

$$\frac{db_o}{dr_o} = \frac{r_o^2 V(r_o)}{2E_o^2 b_o} \frac{dE_o}{dr_o}. \tag{5}$$

When multiple orbiting occurs, $V(r_o)$ must be negative but it is possible for dE_o/dr_o to have either sign; hence db_o/dr_o can be negative in some regions. This possibility seems to have been overlooked in previous discussions of the calculation of the transport cross sections and contradicts the general assertion by Rainwater *et al.* [5] that "the larger of the two impact parameters [at the same E_o] is associated with the larger orbiting radius" and similar but more specific claims by Smith and Munn [3].

We have written a new computer program, PC, that uses the Clenshaw-Curtis method [8, 9, 10, 11, 12] for evaluating well-behaved integrals; this is the method introduced by O'Hara and Smith [4] for evaluating eqs. (1)-(3). Compared to program QVALUES [6, 7], it more carefully handles multiple orbiting for tabulated potentials. Another improvement is that it uses cubic splines to interpolate between the tabulated points on the potential energy curve, rather than a variable-point Lagrangian interpolation. The latter method was highly effective years ago, when only a limited number of potential energy values were tabulated and when the values often had rather large fluctuations in going from one separation to another. The present state of the art, however, is to obtain the potential energies at a large number of separations by CCSD(T) techniques [13] that use large basis sets extrapolated to the complete-basis-set limit, to correct the calculated energies for basis set superposition error, and to include spin-orbit coupling when appropriate. Consequently, it is now better to use cubic-spline interpolation among the tabulated potential values than to use Lagrangian interpolation. Direct comparisons with a recent potential for Cl^- interacting with He [15] has shown that there is excellent agreement (within 1 ppm) in most of the tabulated range of separations, but that the use of a clamped cubic spline gives better results near the ends.

2. Calculation and Comparison

It is possible to compute derivatives of a function by differentiating the Lagrangian interpolation function. For numerical potentials, however, this can lead to unphysical oscillations, due to small errors in the tabulated potential points or to use of a higher-order Lagrangian than is warranted by the precision and number of points. Hence program PC computes potential derivatives in a different manner that is more efficient and more accurate. For separations above the last tabulated value, which is assumed to lie on the known long-range potential, it uses analytical derivatives of this asymptotically-correct function. For separations below the first tabulated value, which must lie on the repulsive wall of the potential, it uses the derivative of an inverse-power potential that matches the first two tabulated points. For intermediate separations, it uses the coefficients computed by the cubic spline to evaluate the first (and, when necessary, second) derivative in a manner [14] similar to that used to evaluate the potential at intermediate separations. The cubic spline is clamped so that it matches the derivatives at separations smaller and larger than the tabulated points. The potential and first derivative are forced by the cubic spline to be smooth and continuous at each tabulated separation, while the second derivative is only continuous.

We have used program PC to obtain the orbiting parameters for Cl^- colliding with He, using an as-yet unpublished potential [15]. The results are compared in Figs. 1-3 with those obtained from the same potential using program QVALUES [7]. Fig. 1 shows that the differences in the interpolations used in PC and QVALUES lead to slight differences in the orbiting energies at small r_o, i.e. for energies near E_c. More significant is the occasional abrupt rise or fall shown by the QVALUES results at certain separations; such abrupt changes are also shown by the b_o values in Fig. 2. Abrupt changes have long been observed [4], and special techniques have been employed in previous programs to make small changes to r_o to avoid them; such techniques are unnecessary in PC. Fig. 3 shows that for this potential for Cl^- in He, E_o is very nearly a single-valued function of b_o. Multiple orbiting occurs only between 6.64×10^{-5} and 7.27×10^{-5} hartree (Fig. 4) and between 0.96×10^{-5} and 1.00×10^{-5} hartree (not shown).

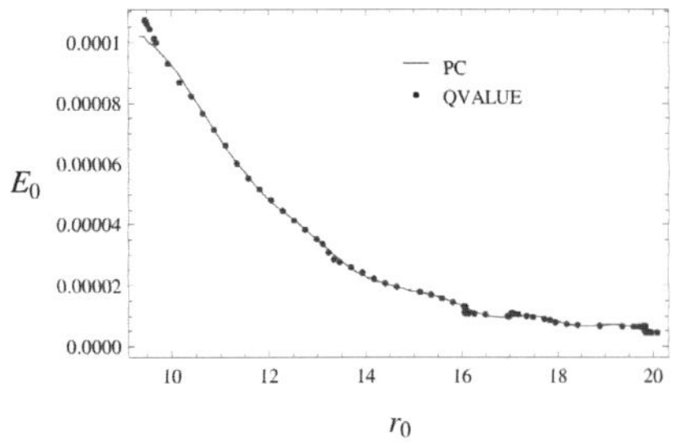

Figure 1. Orbiting energy, E_0 in hartree, as a function of the orbiting separation, r_0 in bohr, for Cl^- collisions with He as computed using the new program, PC, and the older program, QVALUES [7].

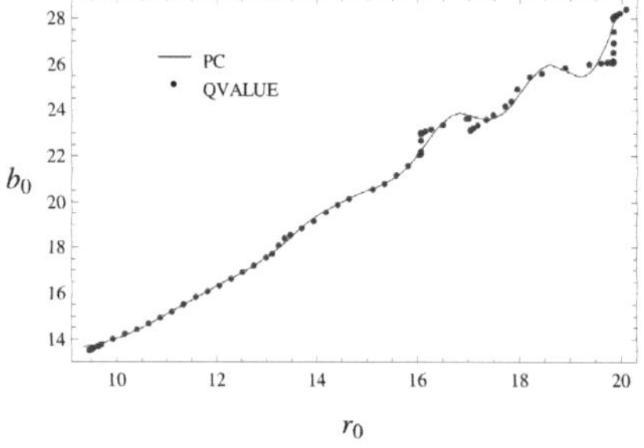

Figure 2. Orbiting impact parameter, b_0 in bohr, as a function of the orbiting separation, r_0 in bohr, for Cl^- collisions with He as computed using the new program, PC, and the older program, QVALUES[7].

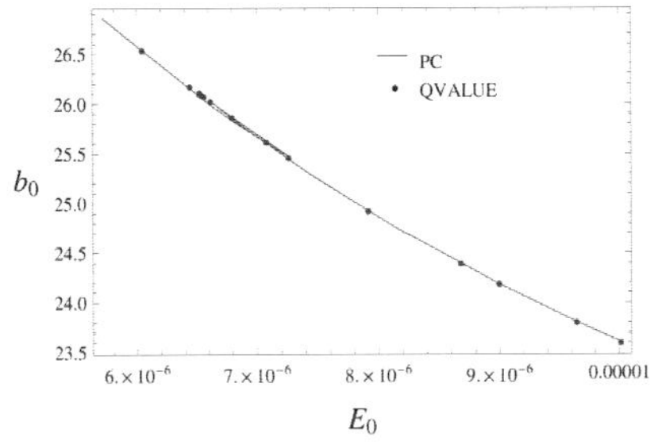

Figure 3. Orbiting impact parameter, b_0 in bohr, as a function of the orbiting energy, E_0 in hartree for Cl^- collisions with He as computed using the new program, PC, and the older program, QVALUES [7].

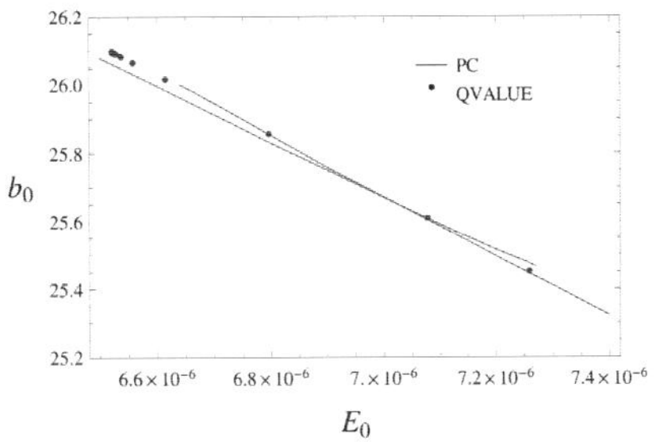

Figure 4. The orbiting parameters obtained by program QVALUES do not show multiple orbiting, following the PC results in multiple orbiting in the energy range between 6.64×10^{-6} to 7.27×10^{-6} hartree.

As can be seen in Fig. 4, multiple orbiting for Cl^- in He was detected by PC but not by QVALUES. This is a result of more careful and more consistent treatment of the potential and its derivatives in PC. We note, however, that the fact that multiple orbiting occurs here only over two small energies regions implies that it makes only a small effect on the cross sections, which involve integrations over wide ranges of impact parameter, and an even smaller effect on the transport properties, which involve energy averages of the cross sections. Hence it is not necessary to redo results obtained over the past 25 years with QVALUES in situations where transport cross sections were obtained to the requested accuracy.

Finally, program PC is designed to work with tabulated potentials for neutral systems, where the long-range potential varies as the inverse-sixth power of the separation, and for atomic ion-atom systems, where it varies as the inverse fourth power. It also works with potentials that vary as the inverse-third power and where there is both a minimum and a long-range maximum; such potentials represent molecular ion-neutral systems where all of the angles are fixed. In further work we will investigate the latter situation more extensively, in the context of the Monchick-Mason approximation [16].

References

[1] Mason E A and McDaniel E W 1988 *Transport Properties of Ions in Gases* (Wiley, New York)
[2] Viehland L A, Janzen A R and Aziz R A 1995 *J. Chem. Phys.* **102** 5444
[3] Smith F J and Munn R J 1964 *J. Chem. Phys.* **41** 3560
[4] O'Hara H and Smith F J 1970 *J. Computat. Phys.* **5** 328
[5] Rainwater J C, Holland P M and Biolsi L 1982 *J. Chem. Phys.* **77** 434
[6] Viehland L A 1982 *Chem. Phys.* **70** 149
[7] Viehland L A 1984 *Chem. Phys.* **85** 291
[8] Clenshaw C W and Curtis A R 1960 *Numer. Math.* **2** 197
[9] Imhof J P 1963 *Numer. Math.* **5** 138
[10] Smith F J 1965 *Numer. Math.* **7** 406
[11] O'Hara H and Smith F J 1971 *Comp. Phys. Comm.* **2** 47
[12] Neufeld P D and Aziz R A 1972 *Comp. Phys. Comm.* **3** 269
[13] Yousef A, Shrestha S, Viehland L A, Lee E P F, Gray B R, Ayles V L, Wright T G and Breckenridge W H 2007 *J. Chem. Phys.* **127** 154309
[14] Press W H, Flannery B P, Teukolsky S A and Vetterling W T 1989 *Numerical Recipes (FORTRAN version)* (Cambridge, University Press)
[15] Lee E P F, Wright T G, Withers C D and Viehland L A to be published
[16] Monchick L and Mason E A 1961 *J. Chem. Phys.* **35** 1676

Journal of Physics: Conference Series **204** (2010) 012008 doi:10.1088/1742-6596/204/1/012008

Mobility of a Li^+ ion attached to 2-butanol in He gas

Shiro Matoba, Tetsuo Koizumi, Takao M. Kojima*,and Hajime Tanuma†

Rikkyo University, Toshima-ku, Tokyo 171-8501, Japan
*RIKEN, Wako, Saitama 351-0198, Japan
†Tokyo Metropolitan University, Hachioji, Tokyo 192-0397, Japan

E-mail: `matoba@rikkyo.ac.jp`

Abstract.
The mobilities of a Li^+ ion attached to 2-butanol in He gas have been measured using a drift tube mass spectrometer at room temperature. We observed a significant reduction in the mobilities over the entire range of E/N.

1. Introduction

Experiments involving collisions of photons, electrons, and molecules with structurally complex molecular ions (such as organic molecular ions) in a gas phase have been performed in recent years to investigate interactions and cross sections. However, few studies have performed collision experiments involving complex molecular ions in the thermal energy region in a gas phase, as it is di cult to reduce the collision energy between an ion and a neutral below 1 eV due to the space-charge e ect in the conventional ion beam method. Cross sections at low collision energies provide invaluable information on the behavior of molecular ions in the environment. The drift tube method is a powerful technique for studying ion–molecule collisions and for measuring gas-phase ion transport parameters, such as mobilities, di usion coe cients, and reaction rate constants, at very low collision energies (i.e., from milielectronvolts to a few electronvolts). Ion mobility in a gas measured by the drift tube method is directly related to the momentum transfer cross section, which is determined by the interaction potential between an ion and a neutral. Therefore, the ion mobility provides not only the momentum transfer cross section at very low collision energies but also detailed information on the interaction potential.

It is necessary to measure the ion mobilities at a wide range of E/N in order to obtain the information of the interaction potential. However, few measurements of the ion mobilities of complex molecular ions have been performed at a wide range of E/N, although a number of studies have been made on structure analyses for complex molecular ions using a drift tube method.

In this paper, we present the mobilities of Li^+ ion attached to 2-butanol($C_4H_{10}O$) in He gas at room temperature. A 2-butanol has two stereo-isomers of (*R*)-(-)-2-butanol and (*S*)-(+)-2-butanol. The mobilities of Li^+(2-butanol) in pure He gas in this work are the basic data for giving the mobility of a chiral molecular ion in a chiral molecular gas with Blanc's law.

2. Experimental setup

We have developed a drift tube for measuring ion mobilities and for observing ion–molecule reactions at room and liquid nitrogen temperature. The details of the experimental setup have been described elsewhere [1, 2]. We brie y mention only the important features and recent updates. The Li^+ attachment ionization is used to produce Li^+(2-butanol) ions that ionize 2-butanol without producing fragment ions. The ions are mass-selected by a quadruple mass selector and are then injected into the drift tube with a kinetic energy of 9 eV. The injected ions are thermalized by collisions with a He bu er gas, and they drift under the in uence of the uniform electric eld in the tube. There are 13 guard rings inside the drift tube that generate the uniform electric eld and six mesh electrodes that function as three electrical shutters for ions. These three shutters are mounted with a distance of 34 mm between the rst and second gates and a distance of 25 mm between the second and third gates. A secondary electron multiplier detects the ejected ions from the drift tube after being analyzed by a Wien lter. The drift velocity is obtained from the slope of the arrival time between each shutter and the detector.

The ionization chamber and main chamber are pumped with turbo molecular pumps of 300 ℓ/s and 1100 ℓ/s, respectively, and a 4-inch oil di usion pump is used to evacuate the detector chamber. The background pressures of the ionization, main, and detector chambers are approximately 5×10^{-3}, 2×10^{-3}, and 1×10^{-4} Pa, respectively, in typical operation with inner pressures of 30 Pa in the ion source and 70 Pa in the drift tube.

3. Results and discussion

The drift velocity v_d is proportional to the reduced electric eld E/N:

$$v_{\rm d} = n_0 K_0 \frac{E}{N}, \tag{1}$$

where N is the number density of the bu er gas molecules, n_0 is the Loschmidt constant and the proportionality constant K_0 is de ned as the reduced mobility of an ion in a gas. K_0 and E/N are in units of $\rm cm^2V^{-1}s^{-1}$ and Td (1 Td = 10^{-17} $\rm Vcm^2$), respectively. On the assumption of an elastic collision and a central force potential, the mean collision energy $\bar{\varepsilon}$ between ions and neutrals in a drift tube and the e ective temperature $T_{\rm eff}$ are given by the Wannier's equation [3]:

$$\frac{3}{2}kT_{\rm eff} \equiv \bar{\varepsilon} = \frac{3}{2}kT + \frac{1}{2}Mv_{\rm d}^2, \tag{2}$$

where k is the Boltzmann constant, T is the bu er gas temperature, and M is the mass of the neutral. Using $T_{\rm eff}$, we can make the comparison between the mobilities measured at di erent gas temperatures. In the classical transport theory of an ion in a gas [4], the momentum transfer cross section at very low collision energies is approximately given by the Langevin cross section, and the mobilities of atomic ions in monoatomic gases are close to the polarization limit $K_{\rm pol}$ in the limit $E/N \to 0$ Td and the bu er gas temperature $T \to 0$ K, which is given by

$$K_{\rm pol} = \frac{13.853}{\sqrt{\alpha\mu}} \quad {\rm cm^2V^{-1}s^{-1}}, \tag{3}$$

where α is the polarizability of the neutral in $\rm A^3$ and μ is the reduced mass between the ion and the neutral in atomic mass units.

The mobilities of Li^+(2-butanol) ions in He gas at room temperature are shown in Figure 1 as a function of (a) the reduced electric eld E/N and (b) the e ective temperature $T_{\rm eff}$.

The mobilities are independent of the bu er gas pressure and are slightly dependent on the reduced electric eld E/N. Notably, the values of the mobilities are signi cantly smaller than the polarization limit $K_{\rm pol}$ = 15.7 $\rm cm^2V^{-1}s^{-1}$ within the measurement range. Previously,

Journal of Physics: Conference Series **204** (2010) 012008 doi:10.1088/1742-6596/204/1/012008

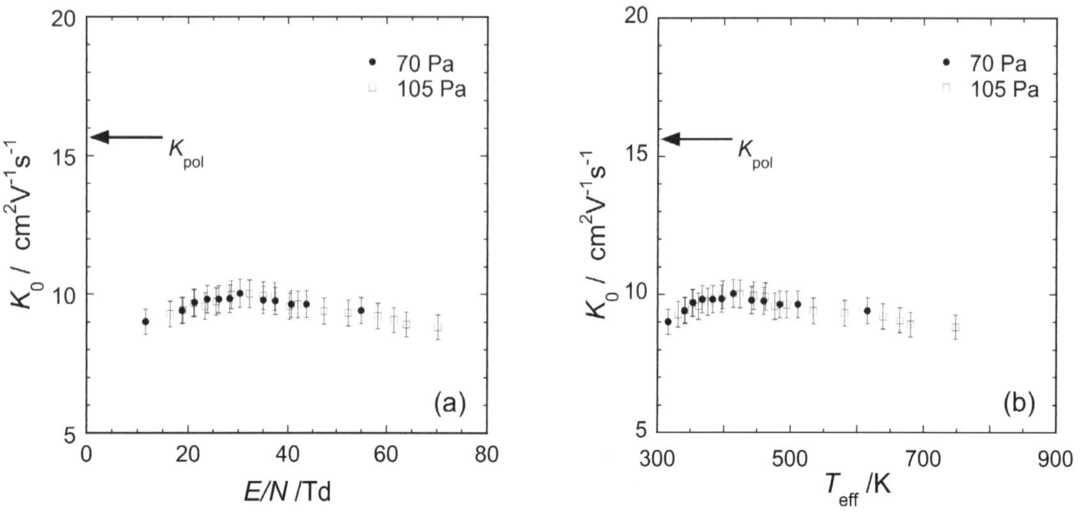

Figure 1. Mobilities of Li^+(2-butanol) ion in He gas at 70 and 105 Pa at 305 K as a function of (a) reduced electric eld E/N and (b) e ective temperature T_{eff}. The arrows indicate the polarization limit K_{pol} at 15.7 $\text{cm}^2\text{V}^{-1}\text{s}^{-1}$.

Hidaka *et al.* and Sandersson *et al.* reported that the mobilities of small molecular ions in He gas at 4.3 K decrease less than K_{pol} at a very low e ective temperature of below 100 K [5, 6, 7]. Ohtsuki attributed the reduction in the mobilities of molecular ions to a Feshbach-like resonance due to the anisotropic potential energy surface between a molecular ion and a He atom predicted by classical trajectory calculations [8]. However, the reduction in the mobilities of the Li^+(2-butanol) ion must be caused by a di erent mechanism because the mobilities do not rise above K_{pol} even at high e ective temperatures. Viehland *et al.* investigated the e ect of the attractive and repulsive terms using the simple (n, 6, 4) model potential [9] given as

$$V(r) = \frac{n\varepsilon_0}{n(3+\gamma) - 12(1+\gamma)} \times \left\{ \frac{12}{n}(1+\gamma)\left(\frac{r_m}{r}\right)^n - 4\gamma\left(\frac{r_m}{r}\right)^6 - 3(1-\gamma)\left(\frac{r_m}{r}\right)^4 \right\}, \quad (4)$$

where n is a parameter of the short-range repulsion energy, ε_0 is the depth of the potential minimum, r_m is the position of the minimum, and γ is a dimensionless parameter that is a measure of the relative strength of the r^{-6} and r^{-4} terms. The r^{-6} term in the ion–atom interaction potential originates from the charge–induced-quadrupole interaction, the permanent-dipole–induced- dipole , and the uctuating dipole–induced-dipole interaction, which is a part of the dispersion interaction. According to the calculations by Viehland *et al.*, the addition of a su ciently high r^{-6} attraction energy can cause the mobilities to fall below the polarization limit K_{pol} at all temperatures. Therefore, there is possibility that the reduction in the mobility of the Li^+(2-butanol) ion is due to the large coe cient of the attractive r^{-6} term in the interaction potential. However, there are other possible mechanisms, such as the e ect of the anisotropic potential, the inelastic scattering, and so on. In order to understand the mechanism that causes the reduction in the mobilities of the Li^+(2-butanol) ion, further studies are required, i.e., the measurements of the mobility at lower temperature and the quantum chemical calculation of the potential energy surface.

4. Summary

We measured the mobilities of Li^+(2-butanol) ions in He gas at room temperature using a drift tube mass spectrometer. The mobilities show lower values than the polarization limit over the entire measurement range.

References

[1] T. Koizumi, T. Tsurugai, and I. Ogawa, 1987 J. Phys. Soc. Jpn. **56**, 17.
[2] T. Koizumi, 1992 J. Phys.B: At. Mol. Opt. Phys. **25**, L335.
[3] G. H. Wannier, 1953 Bell Syst. Tech. J. **32**, 170.
[4] E. A. Mason and E. W. McDaniel, 1988 *Transport Properties of ions in Gases*(New York: Wiley).
[5] J. Sanderson, H. Tanuma, N. Kobayashi, and Y. Kaneko, 1993 J. Phys. B: At. Mol. Opt. Phys. **26**, L465.
[6] J. Sanderson, H. Tanuma, N. Kobayashi, and Y. Kaneko, 1994 J. Phys. B: At. Mol. Opt. Phys. **27**, L433.
[7] H. Hidaka, S. Jinno, H. Tanuma, N. Kobayashi, 2003 J. Phys. B: At. Mol. Opt. Phys. **36**, 1515.
[8] K. Ohtsuki, M. Hananoe, and M. Matsuzawa, 2005 Phys. Rev. Lett. **95**, 213201.
[9] L. A. Viehland and E. A. Mason, 1975 Ann. Phys. **91**, 499.

AUTHOR INDEX

Institute of Physics
Dirac House, Temple Back
Bristol BS1 6BE UK

CURRAN ASSOCIATES INC
proceedings.com

3rd International Workshop on Infrared Plasma Spectroscopy 2008

(IPS 2008)

Journal of Physics: Conference Series Volume 157

Greifswald, Germany
23-25 July 2008

ISBN: 978-1-61738-376-2

IOP Institute of Physics